# What's Happening in the Mathematical Sciences

Volume 8

# What's Happening in the Mathematical Sciences

2010 *Mathematics Subject Classification*:
00A06

ISBN-13: 978-0-8218-4999-6
ISBN-10: 0-8218-4999-9

For additional information and updates on this book, visit
**www.ams.org/bookpages/happening-8**


Printed in the United States of America.
♾ The paper used in this book is acid-free and falls within the guidelines established to ensure permanence and durability.
Visit the AMS home page at
http://www.ams.org/

10 9 8 7 6 5 4 3 2 1 16 15 14 13 12 11

The cover and the frontmatter for this publication were prepared using the Adobe® CS4® suite of software. The articles were prepared using TeX. TeX is a trademark of the American Mathematical Society.

## About the Author

Dana Mackenzie is a freelance mathematics and science writer who lives in Santa Cruz, California. He received his Ph.D. in mathematics from Princeton University in 1983 and taught at Duke University and Kenyon College. In 1993 he won the George Pólya Award for exposition from the Mathematical Association of America. Changing his career path, he studied science journalism at the University of California at Santa Cruz in 1996-97, and has been a freelance writer since then. He has written for such magazines as *Science*, *New Scientist*, *American Scientist*, *SIAM News*, *Discover*, and *Smithsonian*. He was Journalist in Residence at the Mathematical Sciences Research Institute in 2007, and he has written articles and reports for the National Academy of Science, the Simons Foundation, and Simula Research Laboratory in Norway. His first book, *The Big Splat, or How Our Moon Came to Be*, was published by John Wiley & Sons and named an Editor's Choice for the year 2003 by *Booklist*.

## About the Cover Images

**Foreground:** The Hopf fibration arises from a mapping from a 3-dimensional sphere (or "hypersphere") into the ordinary 2-dimensional sphere. Here the 3-dimensional sphere is represented as 3-dimensional space with an extra point at infinity. Every fiber links nontrivially with every other fiber, so that none of them can be pulled apart. Mappings of this type help mathematicians construct and understand the structure of "exotic spheres" and other unusual high-dimensional spaces. (Figure courtesy of Jos Leys.)

**Background:** The standard contact form in 3-dimensional space is illustrated here as a field of planes (the null planes of the contact form). (Figure courtesy of Ken Baker.)

# Contents



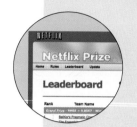

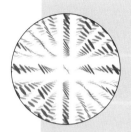

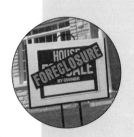

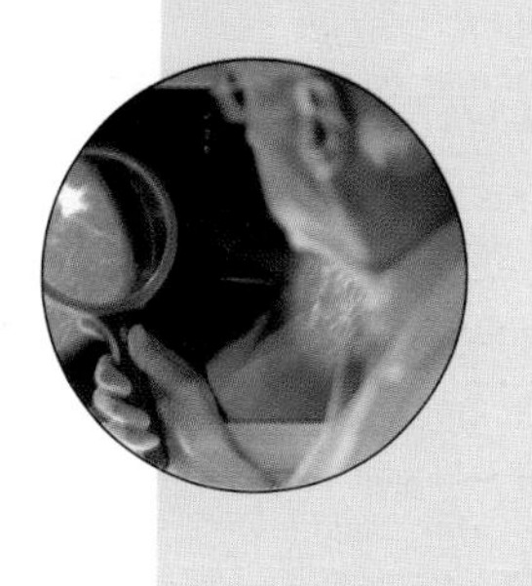

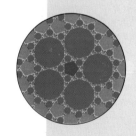

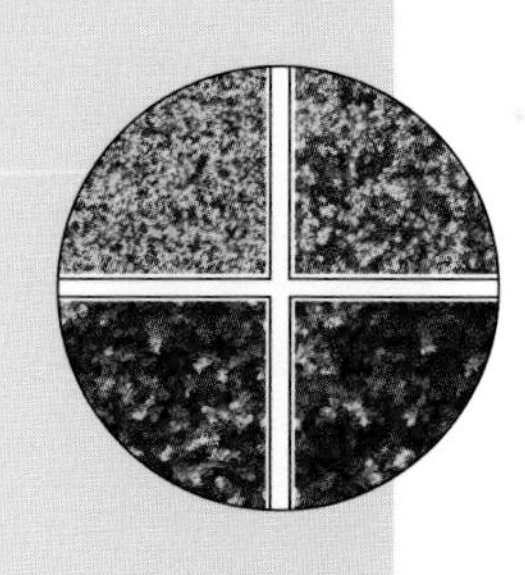

76 **Instant Randomness**

How long does it take to mix milk in a coffee cup, neutrons in an atomic reactor, atoms in a gas, or electron spins in a magnet? In the Ising model of magnetism, mathematicians have calculated a unique cutoff time when the system abruptly goes from essentially unmixed to almost completely mixed. The same behavior is expected for other systems described by the theory of Markov chains.

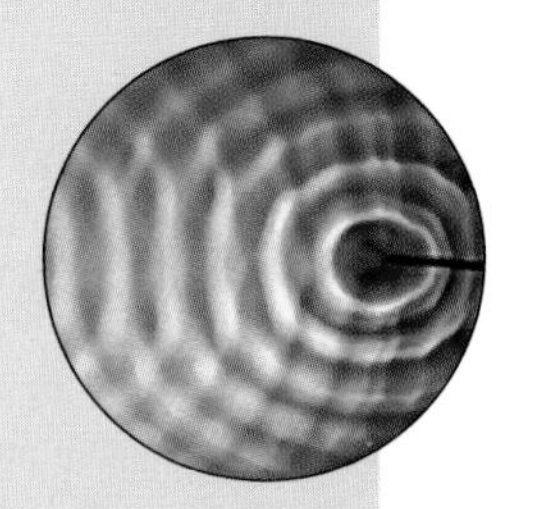

88 **In Search of Quantum Chaos**

A classical billiard ball, on a table with curved, "dispersing" sides, travels on a chaotic trajectory that essentially randomizes the ball's position over the long term. However, at low energies, quantum billiard balls are not chaotic. In a *tour de force* combining pure number theory with physics, mathematicians proved that quantum chaos does emerge at high energies.

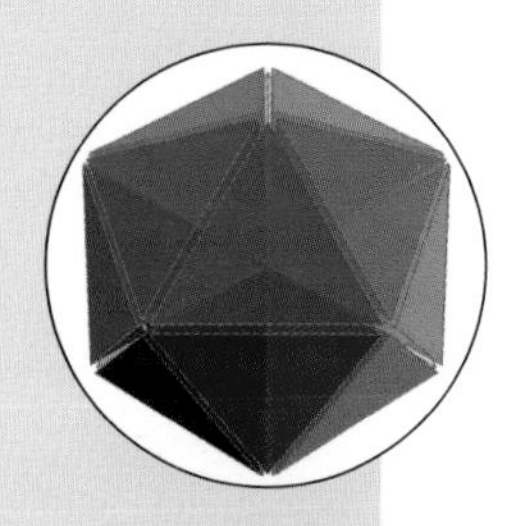

100 **3-D Surprises**

Even in the twenty-first century, mathematics can still reveal new phenomena in ordinary three-dimensional space. Item 1: a stunningly efficient way to pack tetrahedra. Item 2: the Gömböc, a homogeneous body that automatically rights (and wrongs) itself.

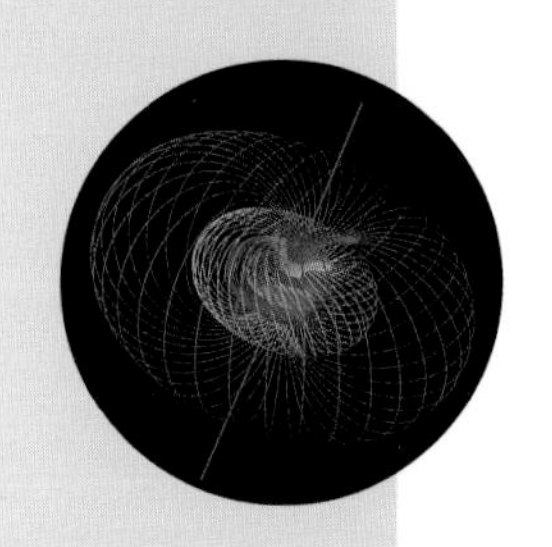

118 **As One Heroic Age Ends, a New One Begins**

John Milnor's discovery of "exotic spheres" in 1956 ushered in a new era of high-dimensional topology, with powerful new tools like framed cobordism theory, stable homotopy theory, and surgery. But one question, called the Kervaire Invariant One problem, remained stubbornly unanswered for more than 40 years, until three topologists found "a shortcut to Mount Everest."

# Introduction

MATHEMATICS ENTERS OUR LIVES in so many ways that we often overlook its presence. We don't think about the number theory behind ATM transactions, or the algebra and geometry that enable rapid internet searches. With each passing year, mathematics has ever more everyday influence, even if it is invisible. Meanwhile, mathematics itself continues to grow, both broader and deeper. This eighth volume of *What's Happening in the Mathematical Sciences* presents some of the remarkable recent mathematical news.

The last few years had current events, both big and small, with mathematical connections. A veil was lifted on the role of mathematics in the economy as various pundits and key players tried to explain the financial crisis brought on by the collapse of major banking and insurance companies. Suddenly, the complex mathematical models that dominate Wall Street were in the news. "Mathematics and the Financial Crisis" looks at whether the models failed or if something else was to blame. A controversy of a different kind erupted with new recommendations for breast cancer screening, which called into question the conventional wisdom on mammograms. "SimPatient" looks at the mathematics that contributed to the reassessment. In a very different vein, mathematics entered the news as the Netflix prize showed how clever programming, clever mathematics, and the judicious combination of multiple algorithms could create a remarkably better method for making recommendations to movie fans, with definite implications for many other possible applications ("Accounting for Taste").

There were new results in familiar areas for mathematics. *Happening* 8 looks at two results related to the dynamics of billiards: one classical but with a twist ("The Ultimate Billiard Shot"), one the quantum counterpart of billiards ("In Search of Quantum Chaos"). Symplectic geometry, which is one way to give geometric formulations of problems in dynamics, saw the solution of a major problem, the Weinstein conjecture, on periodic orbits ("A Brave New Symplectic World"). Mathematicians settled the Kervaire conjecture in topology, which finally answered a nagging question about spheres in higher dimensions ("As One Heroic Age Ends, A New One Begins"). In three-dimensional geometry, there was an amazing race to find the optimum packing of regular tetrahedral shapes. Long thought a dormant topic, a flurry of results showed that there was much we didn't know about the familiar world of three dimensions ("3-D Surprises"). Finally, recent work in probability gave mathematicians much greater insight into randomness, in particular the change from order to disorder ("Instant Randomness").

A lot has been going on in mathematics lately. We are proud to present this eighth volume of *What's Happening in the Mathematical Sciences* with news of some of the outstanding recent progress in mathematics. We hope you enjoy it.

Sergei Gelfand, Publisher
Edward Dunne, Editor

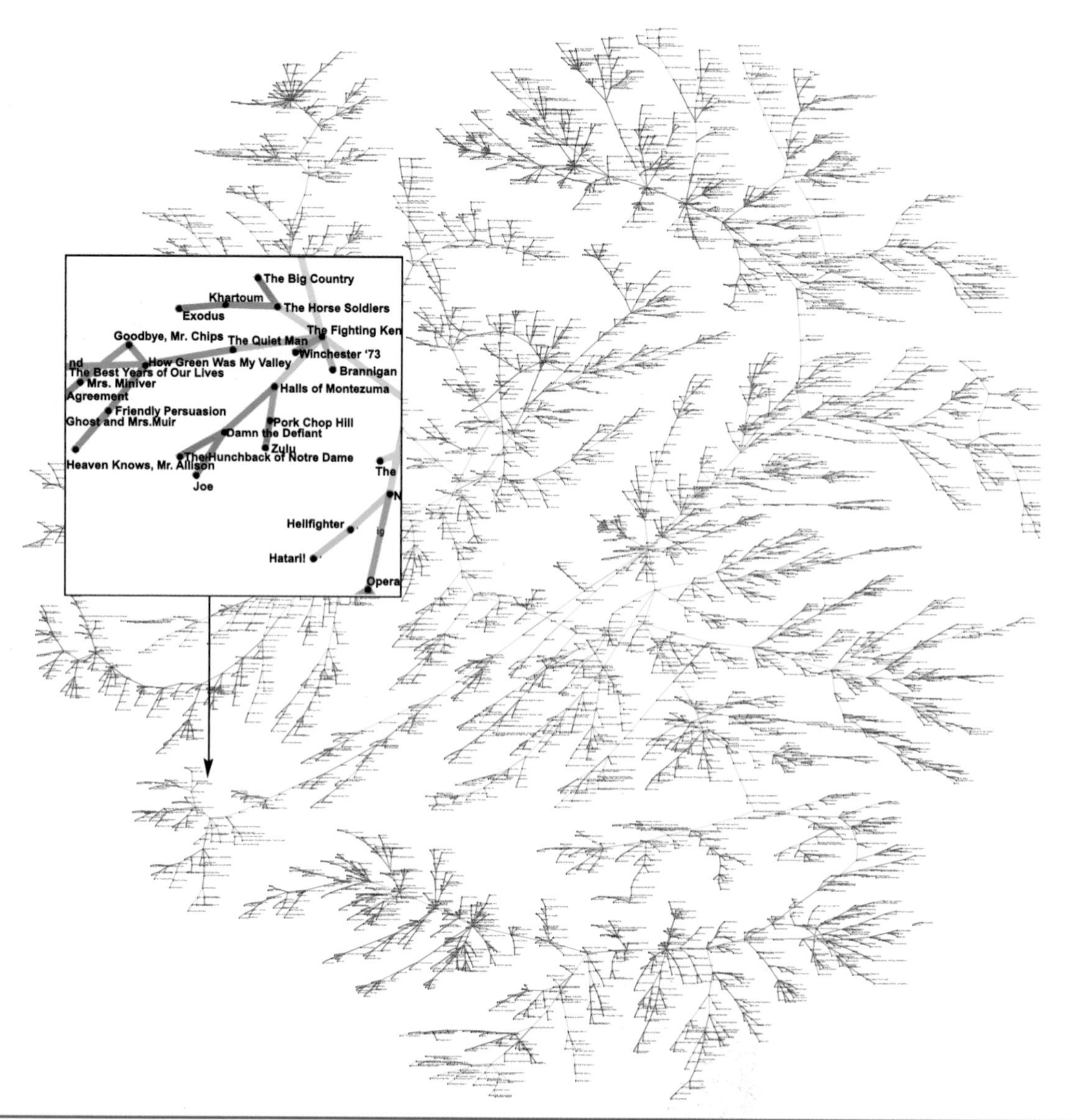


**"Movie Map".** *Nearest-neighbor methods predict the ratings a user will give to a movie by making a map of similarly rated movies. As a stand-alone method they fared unexpectedly poorly in the Netflix Prize, and they were used mostly as a complement to the matrix factorization methods discussed in this article. (Figure courtesy of Christopher Hefele.)*

# Accounting for Taste

FROM ALL OVER THE WORLD THEY CAME: From a cabin in New Zealand, a garage in England, a research laboratory at AT&T in New Jersey. More than fifty thousand people entered what was probably the most publicized competition ever in machine learning, the Netflix Prize. The goal of the competition, organized by the online movie service Netflix, was to develop a computer algorithm that could predict consumers' taste in movies 10 percent better than Netflix's own proprietary program.

In the end, after three years of competition, the award came down to a mad sprint to the finish line, reminiscent of a typical stage of the Tour de France bike race. On September 21, 2009, Netflix declared the winner: A team called BellKor's Pragmatic Chaos.

In fact, the race was a dead heat. The seven-member BellKor team had tied with the second-place team, an enormous collaboration of 33 people called The Ensemble. However, because BellKor's Pragmatic Chaos had submitted its solution 20 minutes before The Ensemble, it claimed the victory.

You might expect the losing team to be bitterly disappointed. But that is where the similarity to a sporting event ends. In fact, the Netflix Prize was a victory for almost everyone involved. Netflix benefited from the publicity, but even more importantly, the contest led to improvements in the company's movie recommendation system. "I know that they are very happy with the improvement," says Charles Elkan of the University of California at San Diego, one of the judges for the competition. "Sometimes people are skeptical and say it was maybe just a PR stunt, but it had a strong technical value."

The contest also led to new jobs and opportunities for many participants. "It didn't give us prize money, but it brought us opportunities and fame," says Gabor Takacs, team captain of The Ensemble. When the competition began, Takacs was a graduate student at the Budapest University of Technology and Economics; now he is a co-founder of Gravity Technologies, a company that specializes in recommendation systems and has 30 employees.

For the relatively new field of computer science called collaborative filtering, the competition provided an unprecedented trial by fire. Some highly touted methods, called nearest-neighbor methods, performed poorly (See **Movie Map** figure). Others, particularly matrix factorization methods, unexpectedly came to the forefront. "Matrix factorization wiped the floor with nearest-neighbor methods," says Lester Mackey, a graduate student at the University of California at Berkeley, who was also a member of The Ensemble.

Finally, for fans of "crowdsourcing," the competition was one of the best examples yet of massively collaborative software design. BellKor's Pragmatic Chaos was the product of a merger between three individual teams (BellKor, Big Chaos,

**Gabor Takacs** *of Gravity, and later the Ensemble. (Photograph courtesy of Gabor Takacs.)*

and Pragmatic Theory). The Ensemble was a merger of several smaller teams who pooled their resources over the last month of the competition (See Figures 1a and 1b). "There has been theoretical research in the machine learning literature showing that ensemble methods are a good strategy for the pure task of predictive accuracy," says Chris Volinsky of AT&T, a member of the winning team. However, neither the organizers nor the participants expected the amount of collaboration that actually occurred.

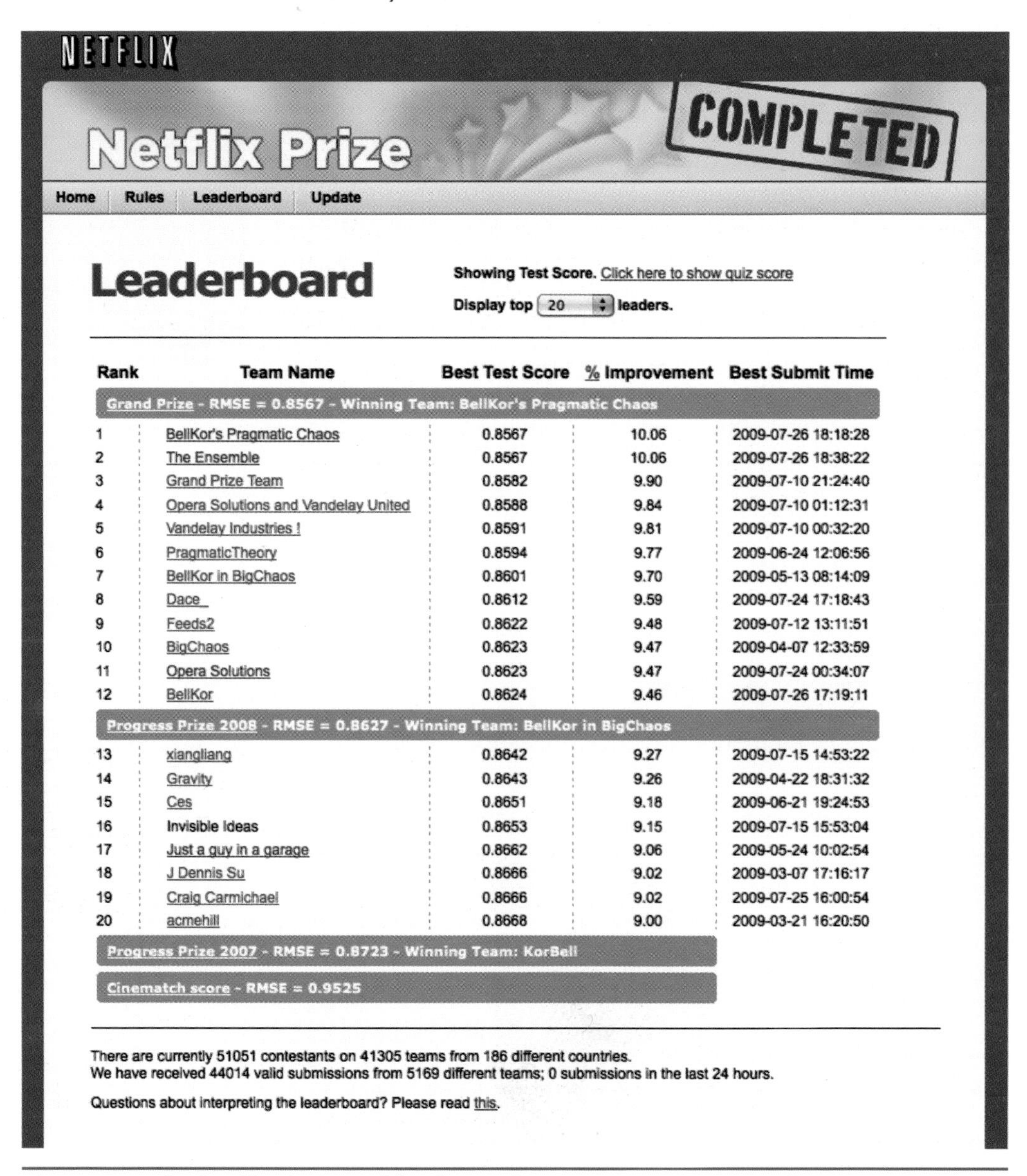
NETFLIX

Netflix Prize

COMPLETED

Home | Rules | Leaderboard | Update

Leaderboard

Showing Test Score. Click here to show quiz score

Display top 20 leaders.

| Rank | Team Name | Best Test Score | % Improvement | Best Submit Time |
|---|---|---|---|---|
| Grand Prize - RMSE = 0.8567 - Winning Team: BellKor's Pragmatic Chaos | | | | |
| 1 | BellKor's Pragmatic Chaos | 0.8567 | 10.06 | 2009-07-26 18:18:28 |
| 2 | The Ensemble | 0.8567 | 10.06 | 2009-07-26 18:38:22 |
| 3 | Grand Prize Team | 0.8582 | 9.90 | 2009-07-10 21:24:40 |
| 4 | Opera Solutions and Vandelay United | 0.8588 | 9.84 | 2009-07-10 01:12:31 |
| 5 | Vandelay Industries ! | 0.8591 | 9.81 | 2009-07-10 00:32:20 |
| 6 | PragmaticTheory | 0.8594 | 9.77 | 2009-06-24 12:06:56 |
| 7 | BellKor in BigChaos | 0.8601 | 9.70 | 2009-05-13 08:14:09 |
| 8 | Dace_ | 0.8612 | 9.59 | 2009-07-24 17:18:43 |
| 9 | Feeds2 | 0.8622 | 9.48 | 2009-07-12 13:11:51 |
| 10 | BigChaos | 0.8623 | 9.47 | 2009-04-07 12:33:59 |
| 11 | Opera Solutions | 0.8623 | 9.47 | 2009-07-24 00:34:07 |
| 12 | BellKor | 0.8624 | 9.46 | 2009-07-26 17:19:11 |
| Progress Prize 2008 - RMSE = 0.8627 - Winning Team: BellKor in BigChaos | | | | |
| 13 | xiangliang | 0.8642 | 9.27 | 2009-07-15 14:53:22 |
| 14 | Gravity | 0.8643 | 9.26 | 2009-04-22 18:31:32 |
| 15 | Ces | 0.8651 | 9.18 | 2009-06-21 19:24:53 |
| 16 | Invisible Ideas | 0.8653 | 9.15 | 2009-07-15 15:53:04 |
| 17 | Just a guy in a garage | 0.8662 | 9.06 | 2009-05-24 10:02:54 |
| 18 | J Dennis Su | 0.8666 | 9.02 | 2009-03-07 17:16:17 |
| 19 | Craig Carmichael | 0.8666 | 9.02 | 2009-07-25 16:00:54 |
| 20 | acmehill | 0.8668 | 9.00 | 2009-03-21 16:20:50 |
| Progress Prize 2007 - RMSE = 0.8723 - Winning Team: KorBell | | | | |
| Cinematch score - RMSE = 0.9525 | | | | |

There are currently 51051 contestants on 41305 teams from 186 different countries.
We have received 44014 valid submissions from 5169 different teams; 0 submissions in the last 24 hours.

Questions about interpreting the leaderboard? Please read this.

**Figure 1a.** *The final leaderboard for the Netflix Prize competition. Netflix users rate movies on a scale of 1 to 5 stars. An "RMSE" of 0.85 means that the algorithm typically misjudged users' ratings by 0.85 stars. Netflix's proprietary algorithm in 2006 (Cinematch) had an RMSE score of 0.95, and teams were challenged to improve on that by 10 percent. Two teams succeeded, and BellKor's Pragmatic Chaos won by virtue of submitting its final algorithm 20 minutes earlier than the Ensemble (18:18 versus 18:38; see final column). (Figure courtesy of Netflix.)*

## A Convoluted History

Imagine a horse race where new horses are constantly entering the race from the side of the track, where individual horses suddenly merge into super-horses and leap to the front, and where nobody actually knows who some of the leading competitors are. That, in a nutshell, describes the convoluted history of the Netflix Prize.

In the beginning, it was a simple idea. The president of Netflix, Reed Hastings, was impressed by the story of the Longitude Prize. In 1714, the English Parliament announced a prize of £20,000 for the first person to develop a method of determining a ship's longitude at sea, to within 30 nautical miles. In addition, there were several intermediate prizes, and eventually more than £100,000 was distributed to various prizewinners. A clockmaker named John Harrison was belatedly recognized as the winner in 1773.

Hastings, too, envisioned a prize that would be awarded in increments, with a $1 million final payoff and $50,000 annual awards for progress to the goal. The task was to improve on the performance of Netflix's own program, called Cinematch, by 10 percent.

At this point some explanation is necessary. How do you predict movie ratings, and what does it mean to improve the accuracy by 10 percent? In fact, the 10 percent figure is a bit misleading; one could make just as good a case for saying that the winners doubled the accuracy of Cinematch... or even better.

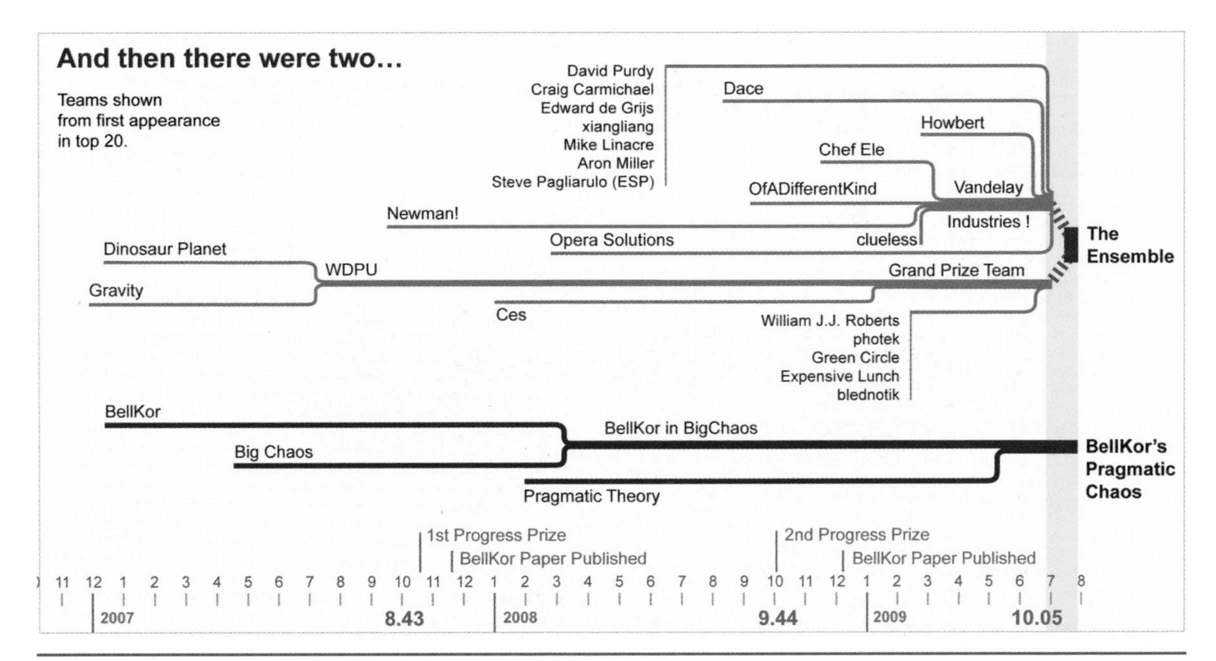


**Figure 1b.** *A striking feature of the competition was the complicated history of mergers between top teams. In fact, most of the teams shown on the leaderboard were absorbed by one of the two leading teams (BellKor's Pragmatic Chaos and The Ensemble). (Reproduced with permission of AT&T Labs, Inc.* 

Netflix offers its customers (all 14 million of them) the opportunity to rate movies, on a scale from one star to five stars (See Figure 2). Though the viewers may do it just for fun—after all, everybody loves to be a critic—for the company these ratings are gold. If you know that a customer liked, for instance, *The Sound of Music*, then there is a good chance he or she will also like *Mary Poppins*, another child-oriented musical starring Julie Andrews. The question is, can you predict how many stars he or she will give it?

At the time, Cinematch was able to predict the customers' ratings to within one star. To be more precise, the root-mean-square error (RMSE) of its predictions was 0.9525; that is how far off a typical prediction is likely to be. On a five-star scale, that may not sound very good. But there is a lot of randomness in the data. If you gave a movie 3 stars today, and forgot what rating you gave it, you might well award it 4 stars next week. In addition to this problem of irreproducibility, an old adage says, "There's no accounting for taste." Could any machine algorithm recreate the irrational likes and dislikes of a human?

Movies Rated Most Often

| Title | # Ratings | Mean Rating |
|---|---|---|
| Miss Congeniality | 227,715 | 3.36 |
| Independence Day | 216,233 | 3.72 |
| The Patriot | 200,490 | 3.78 |
| The Day After Tomorrow | 194,695 | 3.44 |
| Pretty Woman | 190,320 | 3.90 |
| Pirates of the Caribbean | 188,849 | 4.15 |
| The Green Mile | 180,883 | 4.31 |
| Forrest Gump | 180,736 | 4.30 |

Most Active Users

| User ID | # Ratings | Mean Rating |
|---|---|---|
| 305344 | 17,651 | 1.90 |
| 387418 | 17,432 | 1.81 |
| 2439493 | 16,560 | 1.22 |
| 1664010 | 15,811 | 4.26 |
| 2118461 | 14,829 | 4.08 |
| 1461435 | 9,820 | 1.37 |
| 1639792 | 9,764 | 1.33 |
| 1314869 | 9,739 | 2.95 |

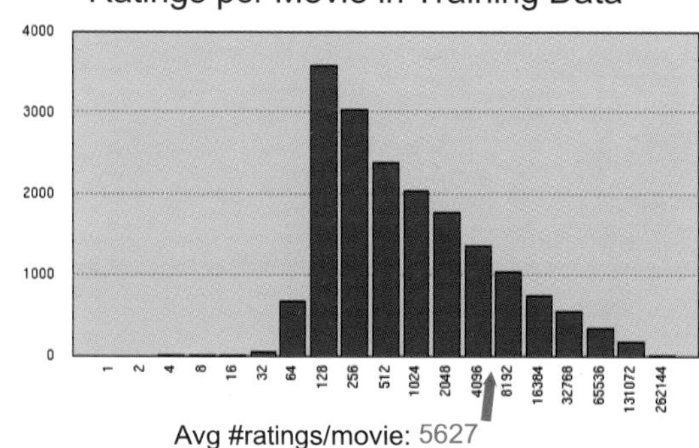


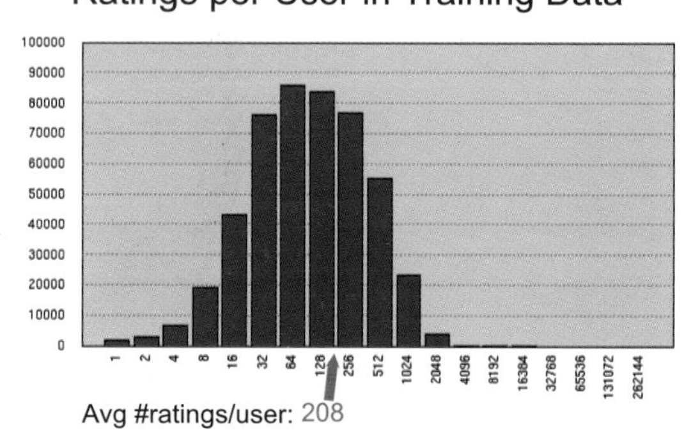


**Figure 2.** *All participants were given access to a database of 100 million ratings provided by 480,000 users for nearly 18,000 movies. Within the database, the number of ratings per user and per movie varied widely. (The most-often rated movie, though far from the most popular one, was Miss Congeniality.) (Reproduced from the AT&T Labs–Research website (www.research.att.com), with permission of AT&T Labs, Inc. All rights reserved.)*

The goal set by the Netflix Prize was an RMSE 10 percent lower than Cinematch's. In other words, the winning program would have to forecast user ratings with a RMSE of 0.8573 or less.

Why 10 percent? Because Cinematch was itself a 10 percent improvement over the simplest possible prediction strategy:

**NAIVE PREDICTION STRATEGY.** To forecast the rating, $r_{ij}$, that user $i$ will give to movie $j$.

1) Compute the average rating that user $i$ gives to all the movies he has rated. (How tough a critic is he?) Call this rating $a_i$.

2) Compute the average rating that movie $j$ gets from all users. (How good a movie is it?) Call this rating $b_j$.

3) Take the average of $a_i$ and $b_j$: $(a_i + b_j)/2$. This is the predicted rating, $\hat{r}_{ij}$. (Here and below, "$r$-hat" denotes predicted ratings, while $r$ denotes actual ratings.)

**John Harrison,** *winner of the Longitude Prize, awarded by the British government in 1773. His chronometers made it possible for sailors to determine their longitude at sea—one of the most vexing technological problems of that era. Netflix CEO Reed Hastings has said that the Longitude Prize was a direct inspiration for the Netflix Prize competition.*

In a sense, this strategy abandons any attempt to understand the interaction between the user and the movie. It predicts that user $i$ will give every three-star movie the same rating, regardless of whether the movie is a thriller, a chick flick, or a cartoon for children. This is the strategy you would use if there truly were "no accounting for taste."

However, in the real world people do have somewhat predictable preferences. As noted above, Cinematch was already 10 percent better than the naive prediction. The goal of the competition, then, was to double this performance, and account for 20 percent of the variance of the real ratings from the ratings predicted by the naive strategy. (Although this may not seem like a gigantic improvement, it would in particular improve the company's ability to rank movies in the order a customer would like them.)

"It was a really lucky stab in the dark," says Volinsky of the target that Netflix set. "The benchmark was interesting enough that it kept us motivated for three years. If the engineers had set a benchmark of 20 percent [improvement], we would have given up long ago. If it had been to improve by 5 percent, we would have gotten it in the first six weeks."

Netflix publicly released a vast quantity of data on user ratings: a table of 100 million ratings of 17,770 movies by 480,189 users. In addition, Netflix withheld two datasets consisting of about 3 million ratings: A quiz dataset and a test dataset. Whenever a team submitted an entry (which they were allowed to do once a day), Netflix would give them a score based on the quiz dataset. The scores were posted on a public leaderboard, so that everybody could see how they compared to the other teams. However, those scores were not official. One reason is that by taking enough guesses at the quiz dataset, teams could either deliberately or inadvertently learn something about that set. If it is inadvertent, this problem is called "overfitting" in the machine learning community—i.e., trying too hard to fit the data that you already have.

Thus, the million-dollar prize and the progress prizes would be based on another, hidden, test dataset. In theory, the scores on the quiz and test datasets should be very similar, because both sets were chosen randomly. In reality, the tiny differences

**Lester Mackey** *of Dinosaur Planet, and later the Ensemble. (Photograph courtesy of Lester Mackey.)*

turned out to be crucial, and the difference between the public leaderboard and the hidden test set led to some confusion in the mass media about who actually won the competition.

A final important rule of the competition was that when any team reached the 10 percent threshold (an RMSE less than 0.8563 on the *public* leaderboard), then a 30-day countdown would begin. All other teams would have that much time to try to catch or surpass the leader—and of course, the leader would also have 30 more days to improve.

The competition began on October 2, 2006. Already in the first year, several teams made major progress toward the 10 percent target. BellKor, which at that point was a three-man group consisting of Volinsky, Bell, and Yehuda Koren (now at Yahoo, Inc.), seemed to have a comfortable lead with just a few days left. Takacs' Gravity (which at that point consisted of four people) was battling Mackey's team, Dinosaur Planet (which at that point had three people) for second place, but a few days before the deadline they decided to combine forces. By averaging their predictions, they found they could score much better on the quiz set than before. On October 1, 2007, the newly merged team, When Gravity and Dinosaurs Unite, caught up with BellKor on the scoreboard.

Both teams feverishly spent the last day trying to improve their scores. On that day, BellKor improved its score from 8.38 percent to 8.43 percent, while When Gravity and Dinosaurs Unite was unable to budge from 8.38 percent. Thus, BellKor had won the first $50,000.

That moment signaled a shift in the strategy of the competition. In the first year, most leading teams had focused on one method of forecasting. But after the merger of Gravity and Dinosaur Planet, it became clear that mergers would be a huge part of any winning strategy. If two teams had prediction methods that somehow captured a different aspect of the interaction between viewer and movie, then a "blended" version of their predictions (either a simple weighted average or some more complicated scheme) would nearly always beat either of the original methods. In year two, BellKor merged with the then #2 team, Big Chaos, forming a new team called BellKor in Big Chaos, and they won the second progress prize with a score 9.44 percent better than Cinematch.

The 10 percent target was now tantalizingly close, but the last fractions of a percent were the hardest. The final breakthrough came when a new team called Pragmatic Theory surged up the leaderboard and passed BellKor in Big Chaos. "Everyone figured that these powerhouses would continue to do battle a while longer, but instead they got together and passed the 10 percent threshold," says Mackey. In reality, the teams had been negotiating a merger for some time, and they had by this time figured out ways to disguise their progress from other teams, by adding random noise to their predictions. In this way, they hoped to catch other teams by surprise when they did merge.

On June 26, 2009, the newly merged team made its appearance on the leaderboard: BellKor's Pragmatic Chaos posted a score 10.05 percent better than Cinematch on the quiz dataset. Many newspapers, confused by the complicated rules, mistakenly reported that BellKor had won. But in reality, the other teams had 30 days to catch up.

And amazingly, they did. Dinosaur Planet and Gravity were, by now, part of an even larger group ambitiously called the Grand Prize Team (GPT). In the final, hectic days of the competitions, a merger was brokered between GPT, Opera Solutions (a private company), and Vandelay Industries (like GPT, a collaboration between many people who had entered the competition separately). The Ensemble was born.

Eager to avoid losing at the wire, as his team had two years earlier, Mackey wanted to keep their existence a secret until the last day. Some of The Ensemble agreed with him, but most of them were afraid of risking everything on a last-minute submission. Thus, the team voted to submit their first entry one day before the 30-day period had elapsed. They now led the competition, with a score of 10.09 to BellKor's 10.08.

Once again, the drama was intense as the final hours and minutes ticked down. Each team would have one more chance to improve its score. BellKor checked in first, 24 minutes before the deadline, with an algorithm that improved its score to 10.09. Just four minutes before the deadline, The Ensemble tweaked another hundredth of a percent out of its algorithm, and its score ticked up to 10.10 percent. "We were in first place!" Mackey says. "We gave a huge collective sigh of relief—we thought we had done the impossible."

But the scores were unofficial, because they were based on the quiz set rather than the test set. On the test set, it turned out that both teams tied at 10.05 percent. The RMSE scores were beyond a photo finish: The Ensemble had an RMSE of 0.856714 on the test set, while BellKor's Pragmatic Chaos had an RMSE of 0.856704. (Remember that RMSE is a measure of predictive error, so a smaller score is better.) However, the rules provided that the RMSE's would be rounded to four digits. Thus, the real margin of victory for BellKor's Pragmatic Chaos turned out to be the 20 minutes by which its solution had preceded The Ensemble's.

## The Mathematical Fallout

The true measure of the success of a scientific prize, of course, is whether it has led to new ideas. And in that respect, the Netflix Prize was a worthy successor to the Longitude Prize.

**The true measure of the success of a scientific prize, of course, is whether it has led to new ideas. And in that respect, the Netflix Prize was a worthy successor to the Longitude Prize.**

First of all, even the presence of such a gigantic dataset was a shot in the arm to the research community on collaborative filtering. Previously, the largest publicly available database on human preferences was one-tenth as large. Some methods that had seemed to work well in the past, such as the nearest-neighbor methods (see the **"Movie Map"**, page 2), failed to scale up well.

**A NEAREST-NEIGHBOR STRATEGY**: To predict the rating, $r_{ij}$, that user $i$ will give to movie $j$.

1) Compute a similarity matrix $s_{jk}$ between movie $j$ and other movies $k$. For example, a statistical approach would use correlation coefficients between user ratings.

2) Compute a function $f_{jk}(r)$ that predicts how a user will rate movie $j$, based on his rating of movie $k$ (for example, by linear regression).

3) Estimate the rating as $\hat{r}_{ij} = \sum_k s_{jk} f_{jk}(r_{ik}) / \sum_k s_{jk}$, where the sum is taken over other movies the user has rated that are comparable to movie $j$.

**Netflix envelope and DVD.** *The red Netflix envelope has become a trusted symbol for movie fans renting films by mail. Few of those fans are aware of the state-of-the-art mathematical methods that Netflix's software uses to generate recommendations.*

In a sense this is an automated version of the example mentioned above: If a user likes one children's musical starring Julie Andrews, he or she will probably like another. Systems that use explicit information, catalogued by humans, have been used successfully in the past (one example is a music-recommending website, Pandora.com). This approach is called content-based filtering. However, content-based filtering was a complete flop in the Netflix Prize: The machines were way better than any human classifier. "We also tried incorporating an extensive set of movie attributes, but this was completely useless," Koren wrote in a comment on the Pragmatic Theory blog. By contrast, Koren, Bell, and Volinsky did use nearest-neighbor strategies in the BellKor algorithm, but as a stand-alone method they were inferior to matrix factorization-based methods.

**A MATRIX FACTORIZATION STRATEGY (CLASSICAL):**

1) Construct a matrix $R$ of all ratings by all users. Note that about 99 percent of the matrix is missing.

2) Fill in the rest of the matrix $R$ some way, for example with 0's or with the naive prediction method.

3) Factor the resulting matrix into a "user feature matrix" $P$ and an "item feature matrix" $Q$, so $R = P^T Q$.

4) Estimate $r_{ij}$ as the $ij$-th element of $P^T Q$, i.e., $\hat{r}_{ij} = \sum_k p_{ki} q_{kj}$. (To avoid overfitting, pick only the $n$ most important features, and perform the sum only from $k = 1$ to $n$.)

The advantage of this method is that a well-known method from linear algebra, called singular value decomposition, can identify the matrices $P$ and $Q$—provided the matrix $R$ is completely known. Unlike the naive method, the matrix factorization strategy is based on a specific model of how users interact with movies. Each movie has some set of features (the genre, the length, the lead actors, the dialogue, etc.), and each user has a set of weights that he or she applies to those features (he likes Westerns, she likes Tom Cruise, etc.) (See Figure 3)

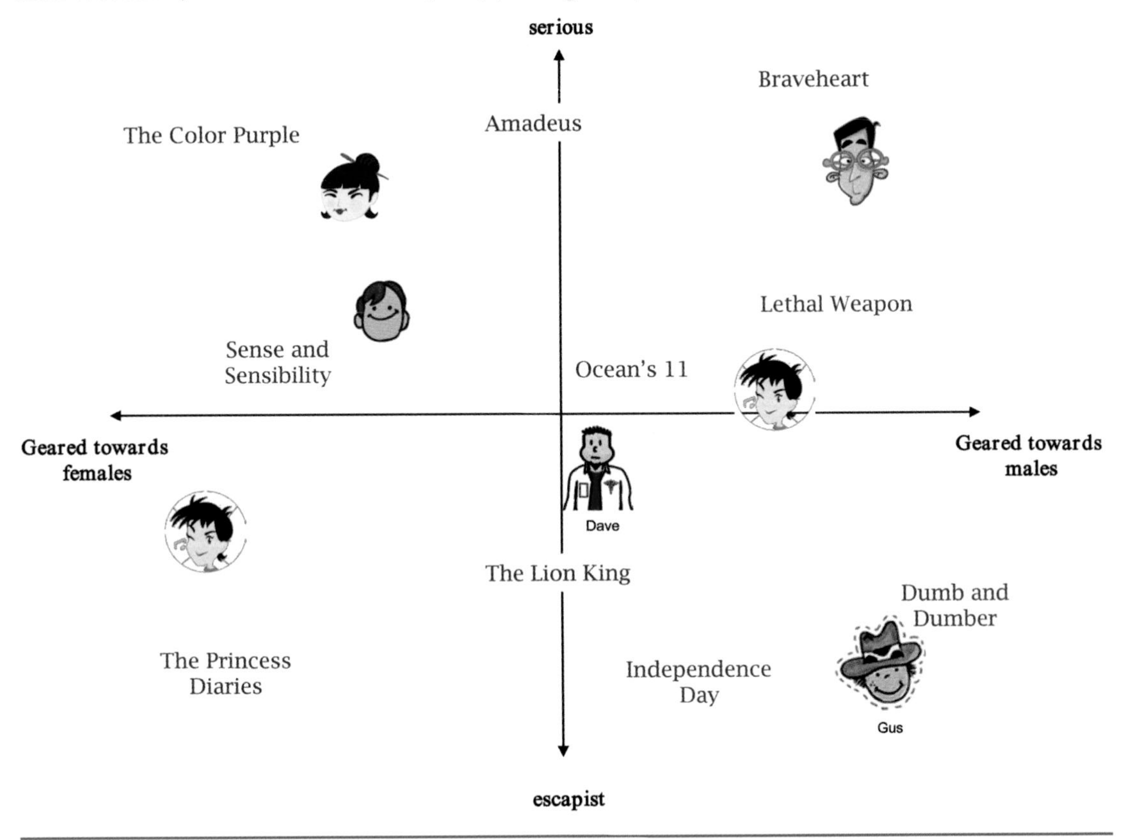


**Figure 3.** *Matrix factorization methods use the training data to infer a vector of features that describe each movie and that describe each user's preferences. In this simplified example, the first feature places "chick flicks" on the left and "guy flicks" on the right, while the second feature places "escapist" movies on the bottom and "serious movies" at the top. The BellKor algorithm identified about 50 features that were most predictive for rating purposes. Unlike the illustrated example, most of them do not correspond to any property that humans can put into words (such as "chick flick" versus "guy flick"). (Figure reproduced from: Figure 2, p. 44, "Matrix Factorization Techniques for Recommender Systems," by Yehuda Koren, Robert Bell, Chris Volinsky. IEEE Computer, August 2009, pp. 42–49. Copyright 2009 IEEE.)*

An important problem with this method is "overfitting," which in this case would arise if you take too large a set of features. It is possible to get perfect equality, $R = P^TQ$, by using linear algebra, but for a number of reasons this is not desirable. First, many of the elements of $R$ were unknown and filled in arbitrarily. We don't want our prediction method to reproduce the arbitrary values that we filled in—we want it to do better than that. Second, one of the main points of the method is to *reduce* the dimensionality of the data. The matrix $R$ has 17 thousand rows and 489 thousand columns. If we can identify a relatively small number of features (in practice, BellKor generally used 50 or so), then the size of the matrices

**However, the ordinary tool chest of linear algebra does not include methods to factor matrices with unknown elements. Thus, [Koren, Bell, and Volinsky] had to devise their own strategy...**

$P$ and $Q$ is much more manageable—17 thousand users (or 489 thousand movies) by 50 features. Finally, Occam's razor suggests keeping the number of features small. If a particular user rated 200 films, you could get a perfect match for his *past* ratings by using 200 or more features—but that would not actually mean you could predict his *future* ratings any better. Because the average user in the database rated about 200 films, the BellKor group wanted to keep the number of features substantially less than 200.

Thus, we should not look for an *exact* factorization $R = P^T Q$ but an *approximate* factorization $R \approx P^T Q$, where for the sake of parsimony the number of features is relatively small (50 or so) and there is an explicit method for minimizing the error. In addition, Koren, Bell and Volinsky did not want to rely on an arbitrary way of temporarily filling in the matrix $R$, as in step 2. They felt it would be much better to keep the unknown ratings as unknowns. However, the ordinary tool chest of linear algebra does not include methods to factor matrices with unknown elements. Thus, they had to devise their own strategy:

**MATRIX FACTORIZATION STRATEGY (WITH REGULARIZATION AND GRADIENT DESCENT):**

1) Define a "regularized squared error" function which, for any pair of feature vectors $\vec{p}_i$ and $\vec{q}_j$, compares their dot product to the *known* ratings $r_{ij}$. For example, $E(\vec{p}_i, \vec{q}_j) = (r_{ij} - \vec{p}_i \cdot \vec{q}_j)^2 + \lambda \|p_i\|^2 + \lambda \|q_j\|^2$.

2) Minimize the sum of $E(\vec{p}_i, \vec{q}_j)$ over all pairs $i$ and $j$ for which $r_{ij}$ is *known*. The problem is too large and nonlinear to solve directly, so take an iterative approach: change $p$ and/or $q$ by a small step in the direction that causes $E$ to decrease the most rapidly. Repeat as often as desired.

3) Eventually, when you are satisfied with all of your user-feature vectors $\vec{p}_i$ and all your item-feature vectors $\vec{q}_j$, you can estimate any *unknown* $r_{ij}$ by the corresponding dot product, $\hat{r}_{ij} = p_i \cdot q_j$.

This is really not just one strategy but a whole class of strategies, in which many variations are possible. In step (1), the error function compares $R$ to the product matrix $P^T Q$, as expected, but it also contains penalties (the two terms with $\lambda$) for including too many features. The strength of the penalty can be controlled by changing the size of $\lambda$. This is called "regularization," and it turned out to be crucial for avoiding overfitting.

In step (2), there are again many options available. In essence, the model is "learning" a row of $P$ and a row of $Q$ at a time. What order should they be learned in? Gravity proceeded chronologically through the known ratings—making an implicit assumption that more recent ratings are more important. BellKor used a method called "alternating least squares," going back and forth between learning $P$ and learning $Q$. The size of the "small steps" in step 2 above, called the "learning rate," is also at the programmer's discretion—or the algorithm itself can be trained to find the best rate. Finally, the number of steps can be chosen at will. "If you only have two hours, you can stop at two hours," Takacs says. "At least the optimization did something."

The first year of the competition identified matrix factorization as the best single approach, for reasons that are still not thoroughly understood. The second year saw much more sophisticated efforts to make use of the time information contained in the Netflix data. For example, certain movies seem to get better with time, while others get worse. If a user rates one movie in a given day, it is likely that he has just watched it; if the user rates 10 or 20 or 5000 in one day (as one user actually did!), he is probably rating them from memory—and that will affect his ratings. While this specific fact may be unique to the Netflix problem, Elkan says, "The general lesson here is that there may be application-specific knowledge that, in composition with application-specific data, allows you to do better."

Other innovations in the second and third years included more subtle approaches to blending. Instead of training two programs separately and then taking a weighted average of their outputs, merged teams could build hybrid models that would train on the database together. The weights themselves would become parameters for the programs to optimize. Also, the large collaborations realized that even methods that did badly on their own could still improve their scores when combined with other methods. Thus both of the leading teams, in the end, blended hundreds of different algorithms, including matrix factorization, nearest neighbors, neural networks, and any other off-the-wall idea that worked.

According to contest judge Elkan, the frantic efforts to squeeze out the fourth decimal point of RMSE were not really the most important part of the contest—although they were certainly the most entertaining. "This demonstrates that contests are never totally realistic," Elkan says. "In the real world, the extra 1 percent of accuracy that you get by averaging across the teams would not be very important. The ideas that were most important came up earlier in the contest." Contestant Takacs agrees with him. "Scientifically, the more interesting things came in the first year. It was a bit less interesting in the second year, and then the third year was all about blending and hacking, improving your score at any cost, and running your ugly algorithm one week for one extra point improvement. That has no value in practice at all."

Indeed, since the end of the competition, the focus of the field has shifted from squeezing out the last point, to digesting the lessons learned and applying recommender systems in new and creative ways. Meanwhile, according to Netflix vice president Steve Swasey, Netflix has incorporated many of the ideas into its recommendation system. "Our customers like it, but they don't realize it," Swasey says. "Most people don't realize there is an enormous amount of computer science behind it; they just see that Netflix recommends this particular movie for them. It's like milk—you know it comes from a cow, but you don't know how it was extracted and pasteurized."

At the time the awards were announced, Netflix announced its plans for a Netflix Prize 2 competition. Since then, however, the plans have been quietly shelved. Perhaps it is for the best, because the first Netflix Prize competition was perfect just as it was. As every movie fan knows, the sequel is almost always worse.

**Indeed, since the end of the competition, the focus of the field has shifted from squeezing out the last point, to digesting the lessons learned and applying recommender systems in new and creative ways.**

***Reeb Orbits.*** *This picture illustrates the standard contact structure on* $\mathbf{R}^3$*, given by the formula* $\lambda = dz - y dx$ *. At every point in space,* $\lambda$ *defines a plane of vectors* $v$ *such that* $\lambda(v) = 0$*. A "Reeb orbit" is a curve that is never tangent to the contact structure, and along which the slopes of the planes remain constant. In this case the Reeb orbits are simply vertical lines. These orbits do not close up. However, the Weinstein Conjecture says that on any compact, 3-dimensional contact manifold (i.e., one with that does not extend out to infinity, as* $\mathbf{R}^3$ *does), closed Reeb orbits always exist. (Figure courtesy of Kenneth L. Baker.)*

# A Brave New Symplectic World

CONSIDER A PENDULUM, SWINGING in a graceful arc from side to side. For centuries clocks depended on its highly repetitive, predictable motion. Over and over again, the pendulum returns to the same location from which it started. As mathematicians would say, it has a periodic orbit.

Now attach the pendulum bob to a spring, instead of a fixed string. The bob now has two degrees of freedom: It can move up and down as well as side-to-side. If the energy is small and the initial conditions are just right, you can still find periodic orbits. But if you stretch the spring too far, or pull the pendulum too far to the side, it starts behaving much more erratically. The stretching of the spring changes the frequency of the pendulum, which starts to lurch from side to side, and up and down, in a chaotic motion. Will the motion eventually repeat itself after a long enough period of time? It is much harder to tell. (See Figure 1, next page.)

In the late 1970s, inspired partly by the springy pendulum, Berkeley mathematician Alan Weinstein formulated a conjecture that has ever since stimulated research in an area of mathematics called symplectic topology. In 2008, Cliff Taubes of Harvard University proved Weinstein's conjecture, showing that for certain kinds of systems with two degrees of freedom, periodic orbits always exist, even if the system is far from equilibrium.

A curious fact about mathematics is that the most important results in certain fields often depend on ideas from outside that field. The fundamental theorem of algebra has never been proven algebraically. The distribution of prime numbers seems to depend on very technical properties of complex functions.

The proof of Weinstein's conjecture follows this pattern. Even though Weinstein posed it as a problem in symplectic topology, Taubes' solution does not stay within that discipline. It borrows heavily on ideas from string theory, developed by physicist Edward Witten, that are still mysterious to mathematicians.

"Seiberg-Witten theory is an amazing thing," says Michael Hutchings of Berkeley, who contributed a key ingredient of the proof called embedded contact homology. "We don't understand why it works or what is going on. Embedded contact homology was my attempt to formulate the content of Seiberg-Witten homology in simple terms, but its definition is crazy. It's not something you would ever think of."

The marriage between physics and symplectic topology makes previously inaccessible problems now seem within reach. Weinstein's conjecture is most likely only the beginning.

**Alan Weinstein.** *(Photo courtesy of Margo Weinstein.)*

### Living in Phase Space

Symplectic topology has its roots in a relatively simple idea discovered by mathematical physicist William Rowan Hamilton in the nineteenth century. Hamilton realized that Newton's laws

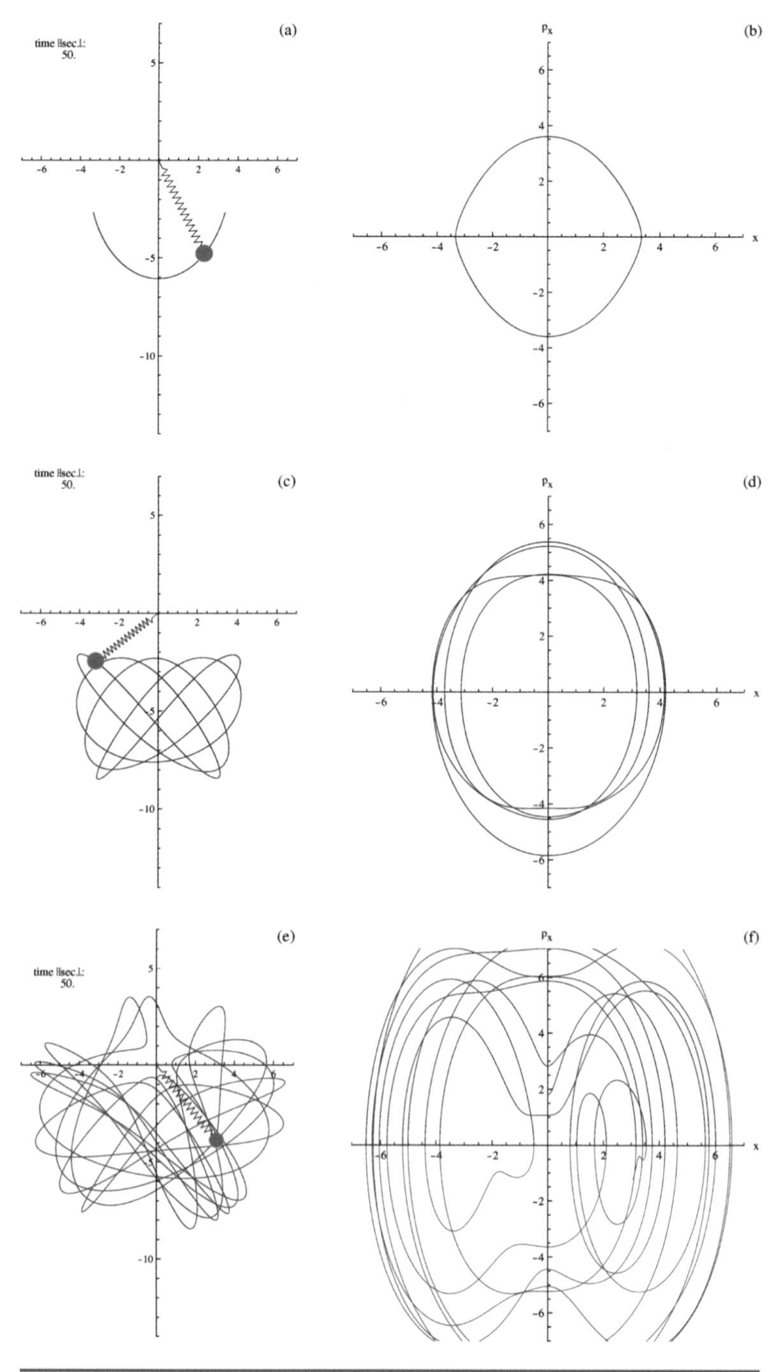


**Figure 1.** *Spring pendulum trajectories plotted in ordinary* $(x, y)$ *space (left) and phase space (right). The full phase space is 4-dimensional, but only the first two coordinates (the pendulum's x-coordinate and x-velocity) are shown here. Note that trajectories (a) and (c) appear to be periodic. However, periodic trajectories are very sensitive to perturbations, and many high-energy trajectories appear chaotic (e). Nevertheless, the Weinstein Conjecture (now proved by Cliff Taubes) guarantees the existence of periodic trajectories for all energy levels. (Graphics created by Arpad Kosa using Mathematica as published by Wolfram Demonstrations (www.wolfram.com).)*

of physics have an elegant geometric interpretation if you think of objects moving not in regular space but in *phase space.* In ordinary space, every object is described by its position. In phase space, every object is described by both its position and velocity. Thus, a body with one degree of freedom (such as a simple pendulum) moves in a two-dimensional phase space. (See Figure 2.) If it moves with two degrees of freedom (like a spring pendulum), its phase space is four-dimensional; and a particle with a full three degrees of freedom has a six-dimensional phase space. Multiple particles can also be considered as a system with a phase space of even more dimensions (but always an even number).

Hamilton discovered that Newton's laws *preserve area in phase space*, that is, if you draw a region in the plane representing a set of possible initial positions and velocities for your particle, and then you let the particle move for a certain length of time according to Newton's laws, then its set of final possible positions and velocities will be a region in the phase plane with the same area, though it may be highly distorted. For example, in Figure 3 (next page), representing the phase space of a pendulum, the red rectangle of starting conditions is mapped to progressively more distorted and stretched regions (also colored red), but the area of each red region is the same. To put it slightly differently, and closer to the way Hamilton thought about it, if you plot all the possible trajectories in the phase plane, they look like the flow lines of an incompressible fluid (in Figure 3, the lines in black). For systems with more degrees of freedom the same thing is true, with area replaced by volume, as Joseph Liouville proved in 1838.

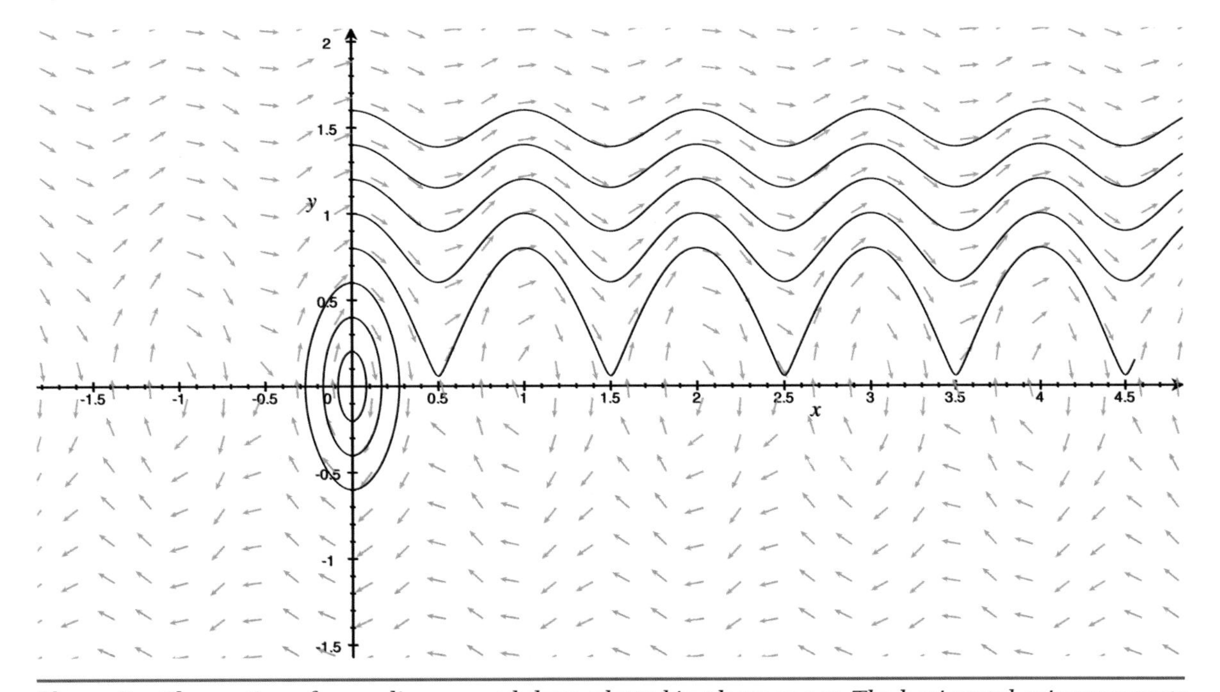


**Figure 2.** *The motion of an ordinary pendulum, plotted in phase space. The horizontal axis represents the angular displacement of the pendulum, and the vertical axis represents its angular velocity. This is a system with a 2-dimensional phase space. All the trajectories are periodic. Even the high-energy trajectories that do not appear periodic are, because when the pendulum swings "over the top," it adds* $2\pi$ *to the angular displacement, but there is no physical difference between angles* $\theta$ *and* $\theta + 2\pi$.

However, Liouville's theorem is not the end of the story. In a Hamiltonian system of particles moving with $n$ degrees of freedom, not only do the trajectories behave like an incompressible fluid, but so do their projections into certain coordinate planes. If you measure only one of the degrees of freedom of your system—say, its motion along the $x$-axis—you will get a 2-dimensional phase plane. If you measure another component of the motion, say along the $y$-axis, you get a second phase plane. Continuing this way until you exhaust all $n$ degrees of freedom, you get $n$ different area measures—and it turns out that their *sum* is preserved by the Hamiltonian flow.

A precise statement of this fact requires an "oriented area" function called a *symplectic form* (or 2-*form*). One characteristic property of this form is that the oriented area of any closed 2-dimensional surface is zero. To close up a surface, you have to give it a front side and a back side, and the oriented areas of the front and back cancel out (one is positive, the other negative).

The subject of symplectic topology deals with generalizations of phase space. These spaces are $2n$-dimensional,

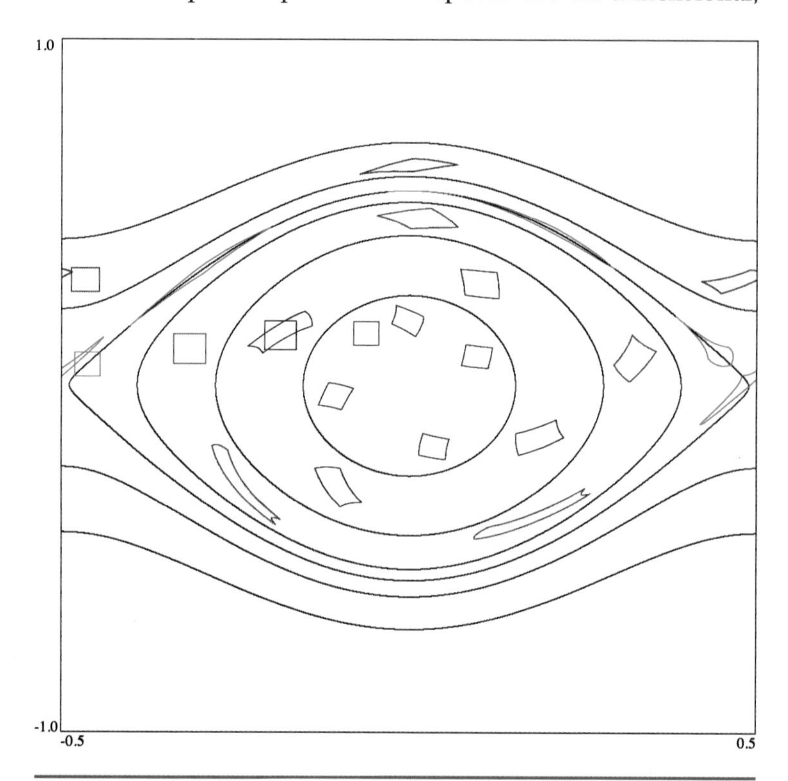


**Figure 3.** *The area-preserving property of Hamiltonian flows, illustrated in the case of the pendulum. The squares represent a patch of initial conditions in the phase plane. For each set of initial conditions, the other regions of the same color represent all possible positions of the pendulum after 1 second, 2 seconds, etc. Although the regions become very highly distorted, their areas are preserved. For example, every red region has the same area as the first (undistorted) red square. (Image generated by Dana Mackenzie using the program StdMap, written by James Meiss.)*

and they have a symplectic form that behaves like the standard symplectic form. It measures the oriented area of 2-dimensional subspaces, it evaluates to zero on small bubbles, and it has an "$n$-th power" that can be used to measure the volume of the entire space. Unlike classical phase space, though, these spaces (called symplectic manifolds) can be curved and have interesting topological features that physicists would call wormholes and mathematicians would call handles.

**. . . two other features give symplectic topology a unique flavor.. . . First, symplectic manifolds are the most natural place to do physics. Second, symplectic topology has strong ties to complex differential geometry.**

## A Smorgasbord of Topologies

At this point, it might be useful to stop and categorize the subtly different flavors of topology and geometry. Topology deals with spaces called manifolds, such as curves and surfaces, that on a small scale resemble Euclidean space. Many mathematicians like to have a little bit of extra structure on their manifolds. For example, if you want to compute derivatives, so that you can make sense of notions like tangency and velocity, then you will study differential topology. The existence of derivatives is a very weak assumption; in fact, it was only in the 1950s that topologists discovered that there is any difference at all between topology and differential topology (see "As One Heroic Age Ends, A New One Begins", page 118).

Some mathematicians like even more structure; they want to be able to use geometric concepts like length and distance. Manifolds with a distance function are called Riemannian, and their study is called differential geometry. The shift in terminology from "topology" to "geometry" reflects a different worldview. Topologists look at the world through lenses in which almost everything looks the same, on a small scale, and only the broadest features can be discriminated. Geometers see a world in which size and shape matter.

Halfway between these extremes lies the field of symplectic topology (as some people call it) or symplectic geometry (as others call it). As discussed above, a symplectic manifold has a symplectic 2-form. This is a geometric notion—for example, it allows you to talk about volumes because Hamiltonian flows have to be volume-preserving. On the other hand, symplectic manifolds are much more flexible than Riemannian manifolds. On a small scale, symplectic manifolds all look the same; their only distinguishing characteristics are large-scale. In that respect, the subject is more akin to topology than to geometry.

But two other features give symplectic topology a unique flavor. First, symplectic manifolds are the most natural place to do physics. Though they are a mathematical abstraction, they could at any time represent the state space of a real physical object—whether it is a pendulum, a vibrating molecule, or a string (the ultimate building block of the universe according to string theory).

Second, symplectic topology has strong ties to complex differential geometry. The symplectic form "almost" gives the space the structure of a complex manifold, i.e., one whose coordinate functions have a real and imaginary part. (For this reason, complex manifolds, like symplectic manifolds, always have even dimension.) In symplectic manifolds, the velocity coordinate can be treated as an imaginary position. The mathematician Hermann Weyl realized this "almost-complex" property when he coined the word "symplectic" in 1938 by

**Mikhail Gromov.** *(Photo courtesy of the Archives of the Mathematisches Forschungsinstitut Oberwolfach.)*

converting the Latin roots "com-" and "-plex" to their Greek equivalents. In 1985 Mikhail Gromov, the 2009 Abel Prize winner, touched off a revolution by exploiting the first-cousin relationship between complex and symplectic geometry.

## Closed Orbits

But before we get to that story, let us talk about the other part of Weinstein's conjecture: The existence of closed orbits. Symplectic geometry may seem hard, but periodic orbits are easy. They are just paths that return to the same place where they started. People have been interested in closed orbits since the dawn of science. The orbits of planets around the sun are closed—or would be, if the planets did not perturb each other's motions. The question of whether these perturbations destroy the periodicity, whether the "music of the spheres" will eventually dissolve into dissonance, has fascinated astronomers from the ancient Greeks to the present day. Likewise, periodic motions are of interest for other physical problems. For example, a simple molecule (such as carbon dioxide) is like a springy pendulum. It has two degrees of freedom, stretching and flexing. It is important to know whether closed orbits of the motion exist, and what their periods are; those periods are intimately related to the molecule's spectrum.

Normally we expect closed orbits to follow a prescribed set of directions, called a vector field, at all times. In physical problems, the vector field often comes from Hamiltonian mechanics; it prescribes the path through phase space that your particle must take. So the question is whether "going with the flow" will eventually take you back home, or doom you to wander forever.

Closed orbit problems do not have to come from physics; one can ask the same question for any vector field on any differentiable manifold. Perhaps the first interesting example is a two-dimensional torus, where we can define a vector field that forces the trajectories to twist around the torus like the stripes on a barber pole. (See Figure 4.) If they twist a rational number of times (say $1\frac{1}{2}$ twists) for each circuit around the torus, then every trajectory will eventually close up. However, if the number of twists is irrational (say $\sqrt{2}$ twists), none of the trajectories close up, and there are no periodic orbits.

This shows that the existence of closed orbits is a very delicate question, where even the tiniest change in the vector field can destroy the orbits. The vector field needs some extra structure if you want to ensure periodic behavior. The condition of being volume-preserving is generally not enough. Weinstein wondered whether preserving a symplectic form, a more stringent requirement, might be sufficient to guarantee periodic orbits. He showed the answer was yes in a special case, where the energy function of the system (called a Hamiltonian) is convex. Later, Paul Rabinowitz generalized the theorem to "star-shaped" Hamiltonians.

These were powerful theorems that took care of many of the problems a physicist would find interesting. But something bothered Weinstein about both of these results. The conditions of being convex or star-shaped are too rigid. They go against the spirit of symplectic topology. Around 1978, he came up with a condition that *was* symplectic in nature. A Hamiltonian system, he conjectured, could be guaranteed to have periodic orbits if the level sets of its energy function had what he called a *contact structure*.

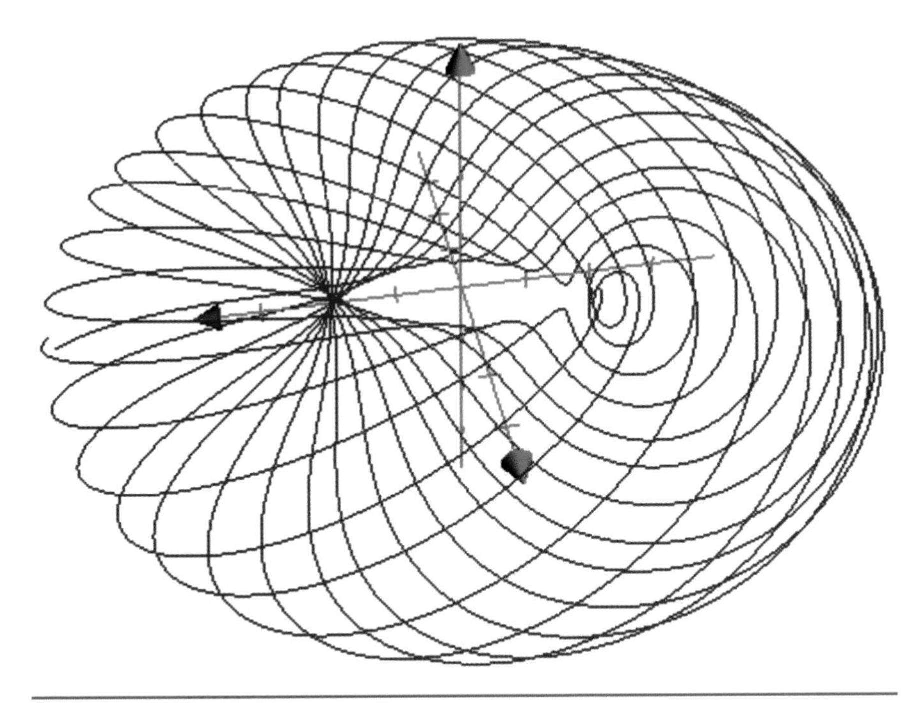

**Figure 4.** *When a trajectory on the torus is tangent to a vector field with irrational slope, the trajectory never closes. In this example, the trajectory "almost closed" after winding around the torus 15 times (see the small gap at left); however, no matter how many times it goes around, it will never close exactly.*

Technically, if $Y$ is a 3-dimensional submanifold of a 4-dimensional symplectic space $X$, then a contact structure is a differential 1-form on $Y$ whose "exterior derivative" is the symplectic 2-form on $X$. The **Reeb Orbits** figure (page 14) gives an idea of what a contact structure looks like in the simplest case, where $Y$ is 3-dimensional Euclidean space and $X$ is 4-dimensional space. The contact structure defines a set of 2-dimensional planes, seen here. Each plane has a unique line of vectors perpendicular to it. The set of all ordered pairs $(y, \mathbf{v}_y)$, where $y$ is any point in $\mathbf{R}^3$ and $\mathbf{v}_y$ is any vector perpendicular to the contact structure at the point $y$, forms a 4-dimensional symplectic manifold. The twisting of the planes, visible in **The Reeb Orbits** figure, is what makes the derivative of the contact structure nonzero.

Intuitively, the importance of a contact structure is that it gives you a way to increase or decrease the energy a little bit without breaking the closed orbits. For example, in the example just cited, different lengths of the vectors $\mathbf{v}_y$ would correspond to different energy level surfaces. You can think of an energy level set as a balloon being inflated; the contact structure tells any point on one energy level set how to move "outward" to the next level set. If the balloon is convex, it is obvious at every point which direction points "outward." But if you have a very convoluted balloon (see Figure 5), it's hard to write down a simple vector field that points outward at every point.

Nevertheless, it can be shown that "almost all" level sets of a Hamiltonian energy function do have a contact structure. This means that periodic orbits are, in a sense, contagious. If you want to find a periodic orbit on a given energy level, you just inflate your balloon a little bit until you find a shape that has a contact structure. Then, if Weinstein's conjecture is true, you can find a periodic orbit on it. Then you deflate the balloon a bit and get a periodic orbit on your original energy level.

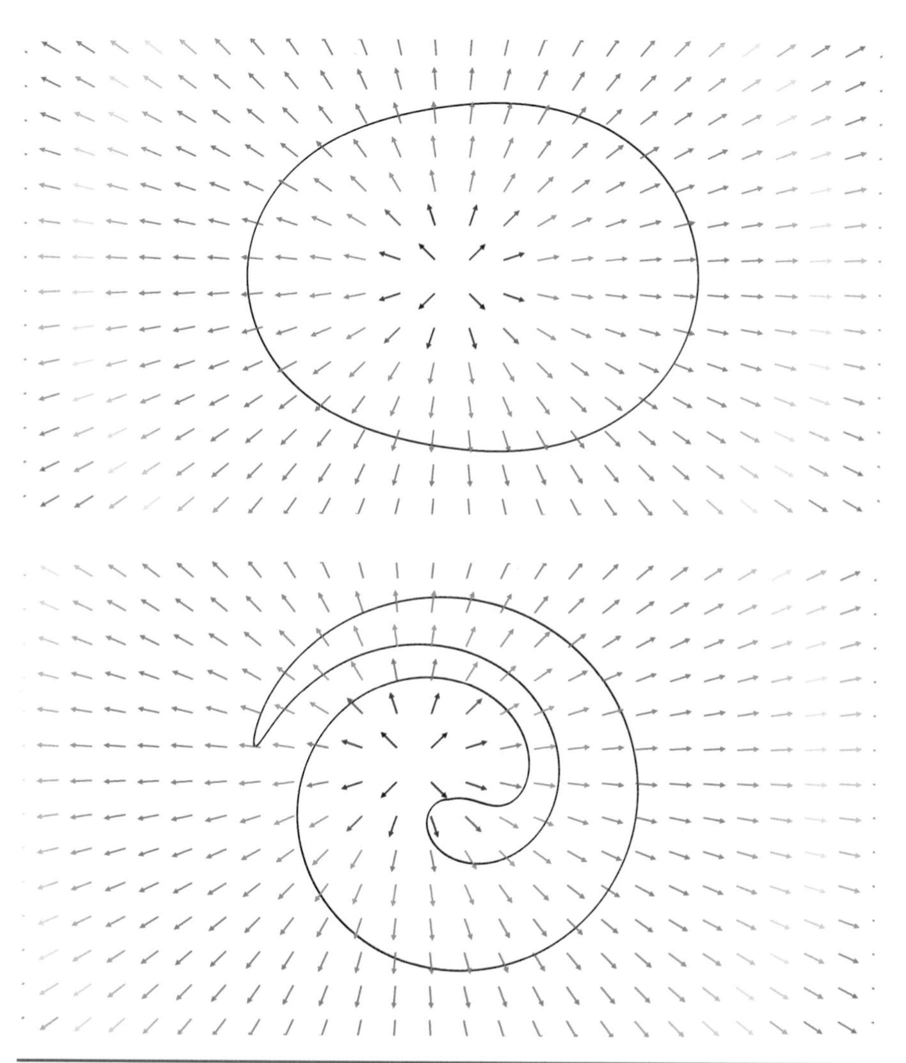

**Figure 5.** *On a convex curve (top), the radial vector field always points outward. However, on a non-convex curve (bottom), the radial field sometimes points outward, and sometimes inward. An outward-pointing vector field does exist, but it would be difficult to write down a simple formula for it. Intuitively, this is the reason that Weinstein's original proof of the existence of periodic orbits applied only to convex Hamiltonians.*

But there was just one problem. Weinstein could not prove that manifolds with a contact structure necessarily have periodic orbits. The problem was *stated* in terms of symplectic topology, but it apparently could not be *solved* by symplectic topology!

## Re-Enter Physics, Stage Left

It would take a strange alchemy of mathematics and physics to break the impasse. The first ingredient was Mikhail Gromov's discovery of pseudoholomorphic curves. This is what you get when you try to force a complex curve (which has two real dimensions but only one complex dimension—that's why it is called a curve) to live inside a symplectic space. It is like a decoy that quacks like a duck and floats like a duck, but isn't a duck.

Gromov showed, for instance, that a Hamiltonian flow can never squeeze a ball into a cylinder of smaller radius. A volume-preserving map can do this easily (think of the ball as being made out of rubber) but a symplectic map cannot. The proof makes use of pseudoholomorphic curves in a clever way, and as Terry Tao has noted in his blog, "it comes down to the fact that one cannot pass a unit ball through a loop of string of length less than $2\pi$."

Vladimir Arnold, a well-known mathematical physicist, nicknamed Gromov's theorem the "symplectic camel theorem," referring to the Biblical statement that "it is easier for a camel to go through the eye of a needle, than for a rich man to enter the kingdom of God." For any object in symplectic space, the cross-sectional area of its fattest embedded disk, a quantity now known as the symplectic capacity, cannot be decreased by a symplectic map. Thus a symplectic camel cannot pass through a needle because its fattest cross section (through one of its humps) cannot be shrunk to the size of the needle.

**Weinstein could not prove that manifolds with a contact structure necessarily have periodic orbits. The problem was *stated* in terms of symplectic topology, but it apparently could not be *solved* by symplectic topology!**

Pseudoholomorphic curves also led to a whole new tool for symplectic topologists, called Floer homology. Andreas Floer was a graduate student who studied with Cliff Taubes at Berkeley in the early 1980s, before returning to Germany and becoming a professor at Ruhr University in Bochum. He died in 1991 at the age of only 34.

When he was still a graduate student, Floer "came into my office one day and said, 'I've figured out how to do Morse theory for the Chern-Simons functional,'" Taubes recalls. "I distinctly remember saying, 'So what?' It was about the stupidest thing anyone could have said, because it's probably one of the most important developments in modern differential geometry."

Morse theory, named after the American mathematician Marston Morse, was one of the earliest methods used to produce topological invariants—quantities that can distinguish one manifold from another. The idea is simple. You define a height function on the manifold $M$. (If the manifold is embedded in Euclidean space, this can really be the height above some coordinate plane.) Each cross section looks about the same as the ones nearby, except at certain isolated points, called critical points, where the topology of the cross section abruptly changes. Each critical point has an index, which simply counts how many directions lead down from that point. (See Figure 6, next page.) The critical points, their degrees, and the number of

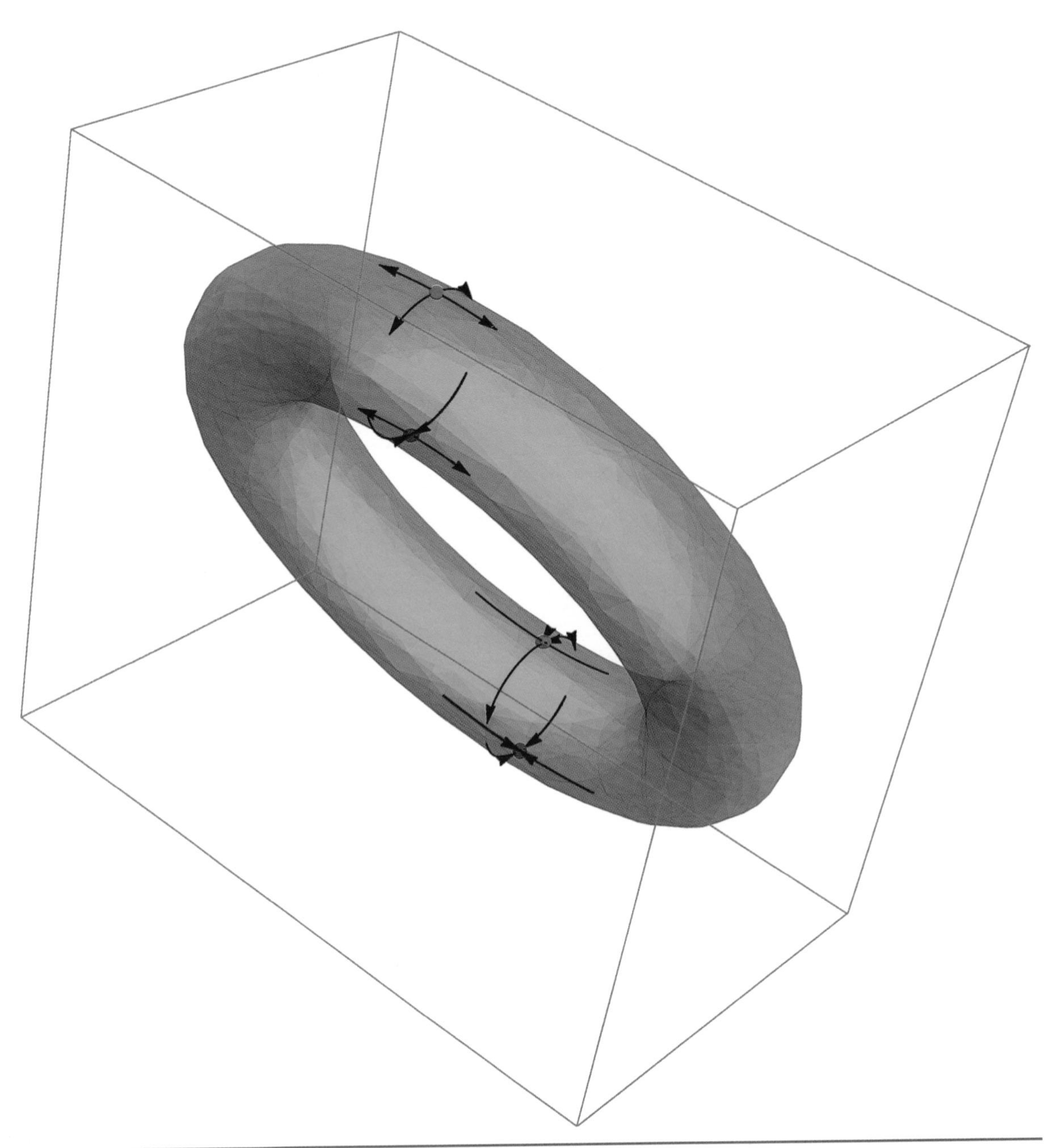

**Figure 6a.** *Morse-Smale-Witten complex on a torus. In classical Morse theory, the topology of a space is detected by means of a height function, whose gradient (the vector field pointing "downward") defines a flow on the manifold. On a torus, the gradient flow has four critical points, shown above. (Figure courtesy of Andrzej Kozlowski. Created with* Mathematica 7*. All rights reserved.)*

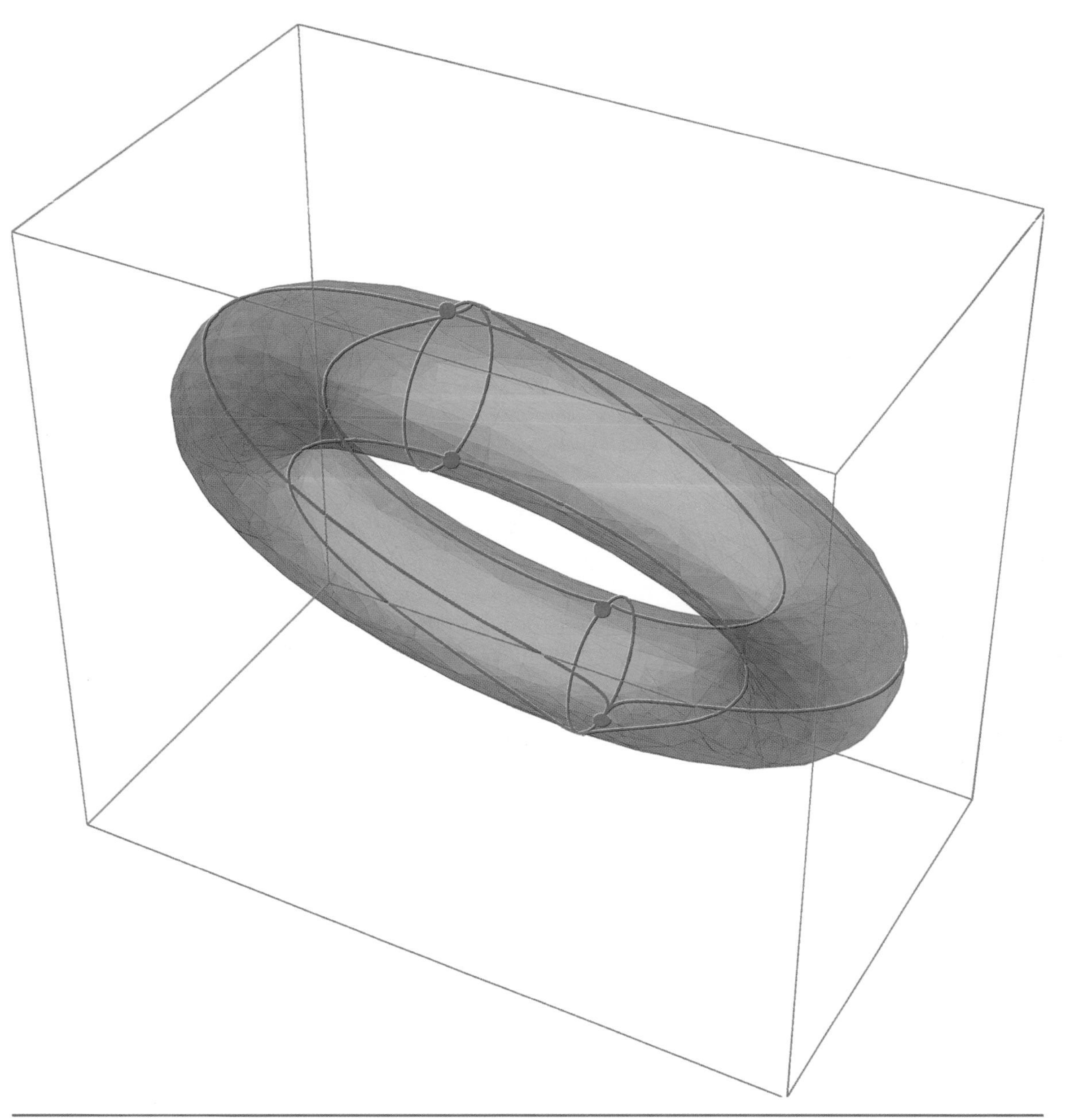

**Figure 6b.** *In Morse-Smale-Witten homology, the special flow lines that join one critical point to another (shown above) form a sort of scaffolding for assembling the manifold. Floer homology does the same thing for the space of loops in a symplectic manifold, with Gromov's pseudoholomorphic curves playing the role of the special flow lines. (Figure courtesy of Andrzej Kozlowski. Created with* Mathematica 7. *All rights reserved.)*

**The Seiberg-Witten equations, and their associated invariants, transformed four-dimensional topology in general, but they transformed symplectic topology the most.**

paths leading from each critical point of index $d$ to each critical point of index $(d - 1)$, provide a basic skeleton for the manifold $M$. Although this skeleton (known as the homology of $M$) does not necessarily suffice to distinguish $M$ from all other manifolds, it contains enough information to answer many topological questions about $M$.

Floer's brilliant, and only temporarily unappreciated, idea was to do the same thing for the space of loops in a symplectic manifold. The "height function" of a loop in symplectic topology is just the symplectic area of any surface that it bounds. (All such surfaces have the same symplectic area. To see why, note that any two surfaces with the same boundary form a bubble, and remember that bubbles always have symplectic area equal to zero. Thus, the area of the first surface minus the area of the second is zero.)

The space of all loops is a vast, infinite-dimensional manifold, and the critical points have infinite index. Thus, at first, it seems that Morse theory (which depends on having a finite set of trajectories between critical points) cannot possibly apply. However, Floer realized that he could define a *relative* index—the difference between the index of one point and the index of another—which is finite (that is, the infinities cancel). The "gradient flow" of the height function defines a two-dimensional tube in $M$, as the loop slides along from one critical embedding to the next. And here is the shocker: this tube is actually a pseudoholomorphic curve. Thus, you have all the ingredients you need to compute the homology of a loop space: critical loops, their indices, and the pseudoholomorphic curves that link them up.

Unfortunately, Floer's original homology was hard to compute. "It's technically a nightmare," says Hutchings. However, after Floer's untimely death a better way came along, which guaranteed that his creation would live on forever. In 1994, Edward Witten and Nathan Seiberg introduced a set of equations that describe the field of a massless magnetic monopole on a manifold. Ironically, magnetic monopoles apparently do not exist in the real world—but in the imaginary world of a symplectic manifold, the equations do have solutions, and the number of solutions is a topological invariant.

The Seiberg-Witten equations, and their associated invariants, transformed four-dimensional topology in general, but they transformed symplectic topology the most. "As soon as the Seiberg-Witten invariants came out, it became clear that you could do Floer homology," says Hutchings. The idea is this: Starting with the solutions to the Seiberg-Witten equations, you can deform them by adding larger and larger multiples of the symplectic form. Eventually the solutions will start looking more and more like pseudoholomorphic curves, which you need for Floer homology. In an 800-page book, Tomasz Mrowka and Peter Kronheimer proved that "Seiberg-Witten Floer homology" makes sense and is nontrivial (in other words, it actually does contain information about the manifold $M$). In the late 1990s, drawing on Mrowka and Kronheimer's work, Taubes showed that the above-described identification between Seiberg-Witten solutions and pseudoholomorphic curves does work, subject to a variety of technical conditions.

Unfortunately, Taubes' theorem does not apply to contact manifolds. It allows us to count 2-dimensional "curves" in 4-dimensional symplectic space, but to solve the Weinstein conjecture we need to count 1-dimensional curves in a 3-dimensional space. (Actually, we merely need to show the count is nonzero.) The recipe for fixing this problem is relatively simple, but actually carrying it out was a ferociously difficult project that took ten years. The idea is to take the 3-dimensional manifold $M$ and "thicken" it by taking its Cartesian product with a line. (For example, if you thicken a circle in this way, you get a cylinder.) The new manifold $M \times \mathbf{R}$ can be given a symplectic form consistent with the contact structure. The product manifold has a Seiberg-Witten Floer homology, which can be translated back down to the original manifold $M$ because the extra factor of $\mathbf{R}$ does not affect the topology (aside from changing the dimensions by 1).

But there are technical issues to overcome. Magnetic monopoles on $M \times \mathbf{R}$ do not necessarily correspond to thickened orbits of the Hamiltonian flow. They do not necessarily have cylindrical symmetry; as Mrowka says, "They are subtle global objects not preferring to be localized in any particular direction." But once again, Taubes showed that when they are deformed by a larger and larger multiple of the symplectic form, they acquire a preferred direction and look more and more like a cylinder over a closed orbit. The biggest technical difficulty for Taubes was to prove that the length of these closed orbits would not approach infinity—so that the Hamiltonian system would in fact return to its starting place after a finite time.

The conclusion, then, is that a homology theory for $M$, generated by closed orbits of the Hamiltonian vector field (this is the "embedded contact homology" defined by Hutchings), could be computed from the Seiberg-Witten Floer homology for $M \times \mathbf{R}$. Because of Mrowka and Kronheimer's work, the latter homology is nontrivial. Thus, closed orbits must exist.

It remains unknown whether Weinstein's conjecture is true for higher-dimensional contact manifolds. Thus, only Hamiltonian systems with two degrees of freedom (like the spring pendulum or the carbon dioxide molecule) are covered by Taubes' theorem. The reason that the proof cannot be generalized is simple—no one knows how to define something like a magnetic monopole on a space of more than four dimensions. There is no Seiberg-Witten equation in six-dimensional space, or at least no one knows what its analogue might be.

On a more positive note, it is clear that the machinery of Floer homology and its relatives can be used to prove much more than Weinstein conjectured. In fact, Weinstein's conjecture pops out as an almost embarrassingly simple consequence of Taubes' work. It is like proving that at least one person lives on Manhattan: It's true, but it is a bit of an understatement. In most cases, Hutchings says, there are many closed orbits—in fact, probably infinitely many. So it will be a challenge for future geometers to see if they can extract more information about the "inhabitants" of their symplectic Manhattan. As Miranda exclaims in Shakespeare's play *The Tempest*: "O brave new world, that has such people in it!"

**It is like proving that at least one person lives on Manhattan: It's true, but it is a bit of an understatement.**

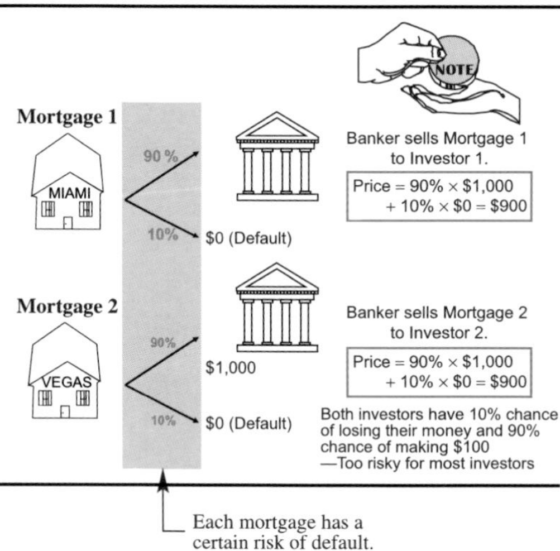


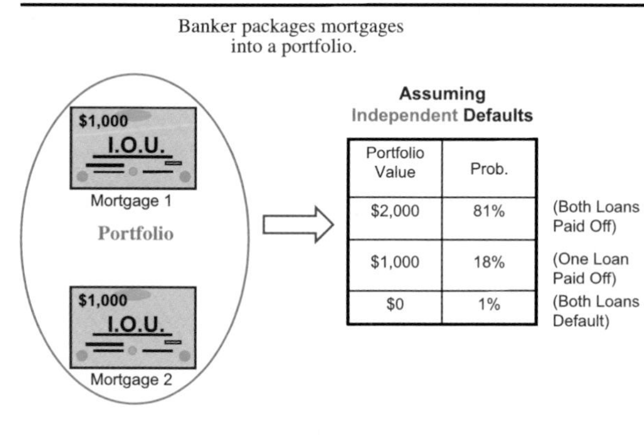


| Portfolio Value | Prob. | |
|---|---|---|
| $2,000 | 81% | (Both Loans Paid Off) |
| $1,000 | 18% | (One Loan Paid Off) |
| $0 | 1% | (Both Loans Default) |

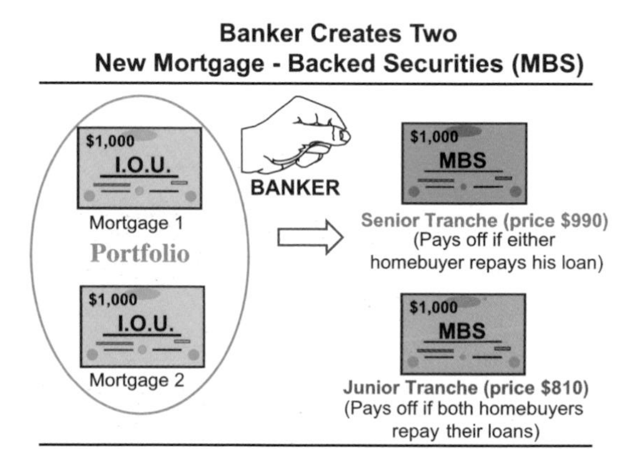


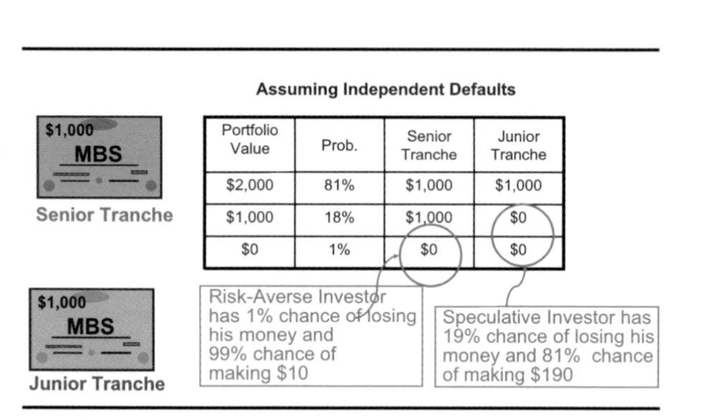


| Portfolio Value | Prob. | Senior Tranche | Junior Tranche |
|---|---|---|---|
| $2,000 | 81% | $1,000 | $1,000 |
| $1,000 | 18% | $1,000 | $0 |
| $0 | 1% | $0 | $0 |



**Securitization.** *The housing boom in the mid-2000s was fueled in part by the creation of mortgage-backed securities, which gave investors the impression that they could invest at lower risk than before. (a) Before the 1990s, investors could purchase mortgages from a bank and make a profit if the borrower paid back the loan. Such investments were considered high-risk. (b) "Mortgage-backed securities" offer differing levels of risk, called tranches. The senior tranches were considered to be very low-risk investments, even by credit rating agencies. This assessment depends on the assumption that the underlying loans have independent probabilities of default.*

# Mathematics and the Financial Crisis

SEPTEMBER 15, 2008, WAS THE TRIPLE WITCHING HOUR of a crisis that had been brewing for years, and whose repercussions are still being felt years later. In the wee hours of that Monday morning, Lehman Brothers, an investment bank, filed for bankruptcy (which would become the largest bankruptcy in U.S. history). Later that day, Bank of America bought Merrill Lynch, another famous investment bank that was on the verge of failure. And three credit rating agencies downgraded American International Group (AIG), the world's largest insurance company, placing it also in imminent danger of collapse. The next day, the U.S. government announced an $85 billion bailout of AIG, a move that was unprecedented in U.S. history and in direct contradiction to the Federal Reserve Bank's own promises two days earlier. The phrase "too big to fail" entered the American lexicon.

A collapse of the world's financial markets was averted, but the consequences were profound nevertheless. Stock markets took a beating over the next few months, and the U.S. entered its most serious recession since the 1930s. More government bailouts ensued, and U.S. taxpayers found themselves as the unwilling creditors of hundreds of billions of dollars of debts that may never be repaid.

The list of possible scapegoats for the financial crisis is long and varied, but *Wired* magazine in 2009 pointed the finger at an unlikely culprit: Mathematics. In an article called, "Recipe for Disaster: The Formula that Killed Wall Street," financial blogger Felix Salmon wrote: "[the] Gaussian copula formula will go down in history as instrumental in causing the unfathomable losses that brought the world financial system to its knees." In fact, virtually every transaction that got AIG into trouble was made possible by mathematical models that have been developed over the last 40 years.

**Timeline of Financial Crisis**

| 6/2003 | 4Q 2005 | 2006 | 4/3/2007 | 7/31/2007 | 3Q 2007 |
|---|---|---|---|---|---|
| Federal Funds Rate lowered to 1 percent, a 45-year low. | US residential housing index drops—the beginning of the end of the "housing bubble." | 20 percent of all mortgage loans are subprime. | New Century Financial, largest subprime lender in US, files for bankruptcy. | Two Bear Stearns hedge funds, which specialized in subprime debt investments, file for bankruptcy. | US GDP grows by 2.3 percent (annual rate). |

*Wired* was not the only one blaming the quantitative analysts, or "quants," for Wall Street's troubles. Nassim Taleb, co-director of New York University's Research Center for Risk Engineering, wrote two best-sellers, *Blinded by Randomness* and *The Black Swan*, which criticized the mathematical models used throughout the financial industry (including the Gaussian copula). Taleb argued that these models systematically underestimated or ignored the importance of rare events or "black swans"—events like liquidity crises, or precipitous drops in the stock market, that are far from the norm. "I think physicists should go back to the physics department and leave Wall Street alone," Taleb told the *New York Times* (a reference to the fact that many of the "quants" have Ph.D.'s in physics).

But mathematical models also have their defenders, such as Andrew Lo, a professor of financial engineering at Massachusetts Institute of Technology. "Blaming quantitative models for the crisis seems particularly perverse, and akin to blaming arithmetic and the real number system for accounting fraud," he wrote in *Journal of Investment Management* in 2010. Lo argued that the blame for the misuse of quantitative models still lies with humans. Risk managers, he said, were guilty of "physics envy." They had turned a blind eye to the assumptions inherent in the models and the uncertainties that still lurked behind them. He also criticized the upper management who did not question the quantitative models—or even worse, actively encouraged their misuse as cover for risky business practices. Instead of "killing all the quants," he argued that Wall Street should hire *more* people with mathematical literacy.

Did mathematics really bring on the financial crisis? What is the Gaussian copula formula? How did mathematics become so powerful on Wall Street, and what can we do about it? Those are questions with a long and tangled history.

## The Black-Scholes Formula

There are many places to begin the story, but a good place to start is with the Black-Scholes Formula, the first great triumph of quantitative methods on Wall Street, which in many ways was the paradigm for everything that followed.

| 10/11/2007 | 10/2007–12/2007 | 4Q 2007 | 1/11/2008 | 2/13/2008 | 3/13/2008 |
|---|---|---|---|---|---|
| Number of foreclosures in US housing market has doubled in the past year. | Citigroup, Merrill Lynch, Bank of America, UBS, Barclays all announce losses due to write-offs or "write-downs" of bad debt. | US GDP grows by 2.9 percent (annual rate). | Bank of America announces purchase of Countrywide Financial , a troubled subprime lender , for $4 billion. | Economic Stimulus Act of 2008 signed. | Bear Stearns announces its liquid assets have dropped by 88 percent. |

The year was 1973. Bell-bottom pants were in, Richard Nixon was President, and a new kind of financial security made its debut: the option contract. Although options had been traded over the counter before, that was the year the Chicago Board Options Exchange was founded, making their trade much easier.

An option is a contract that allows the holder to buy or to sell a security (such as a stock), at a later date (the maturity date), at a fixed price (the strike price). Options come in two flavors, the "call" (an option to buy) and the "put" (an option to sell). The value of a basic call on its maturity date is illustrated in Figure 1(a), next page. If the stock is currently trading at less than the strike price, then the option to buy is worthless; you can buy the stock on the open market for less. On the other hand, if the stock is trading at a higher value, then the value of the option is the difference between the current trading price and the strike price—because that is how much you save by exercising your option. The value of a put option is exactly the opposite, as shown in Figure 1(b), next page: it has zero value if the stock is currently trading above the strike price, but positive value if the stock is trading below. Thus, from the investor's point of view, a call pays off if the stock price moves up, while a put pays off if the stock price moves down.

The question that Fischer Black and Myron Scholes (and simultaneously, Robert Merton) answered was this: How can you estimate the value of a put or call option *before* its maturity date? If you can tell how likely the option is to be "in the money" or "out of the money," just as a result of normal chance variations in the stock price, then you have the basis for setting a fair price for the option.

Black and Scholes constructed a simple mathematical model in which the asset price $S$ satisfies a stochastic differential equation,

$$\frac{dS}{S} = \sigma dX + \mu dt$$

Here $dS/S$ represents the proportional change in the asset's value over the time interval $dt$. It contains a deterministic part ($\mu dt$) that is just the amount a similar investment in a risk-free

**3/16/2008**

JP Morgan Chase announces purchase of Bear Stearns for $2 per share (compared to $30 market price of stock two days earlier).

**3/24/2008**

Federal Reserve Bank of New York provides financing for JP Morgan Chase purchase of Bear Stearns; JPMC sweetens offer to $10 per share.

**1Q 2008**

US GDP decreases by 0.7 percent (annual rate).

**5/8/2008**

US Treasury Secretary Henry Paulson says that we are closer to the end of the credit crisis than the beginning.

**5/2008**

US home prices drop by 15 percent for the second month in a row.

**5/2008–6/2008**

Moody's and Standard and Poor's admit to errors in computer models used to rate "complex debt products."

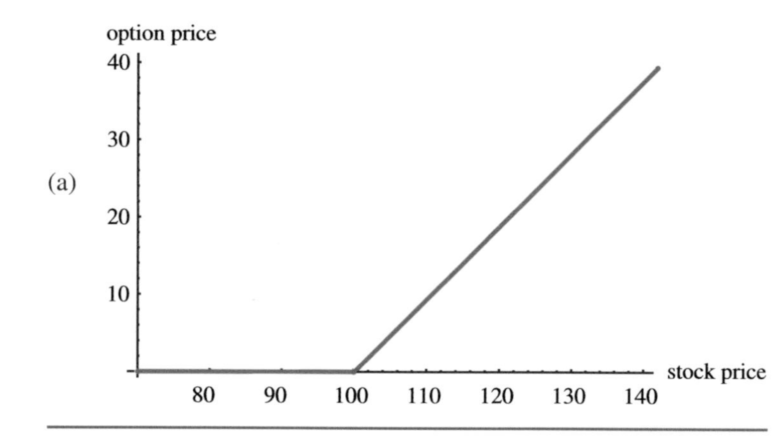


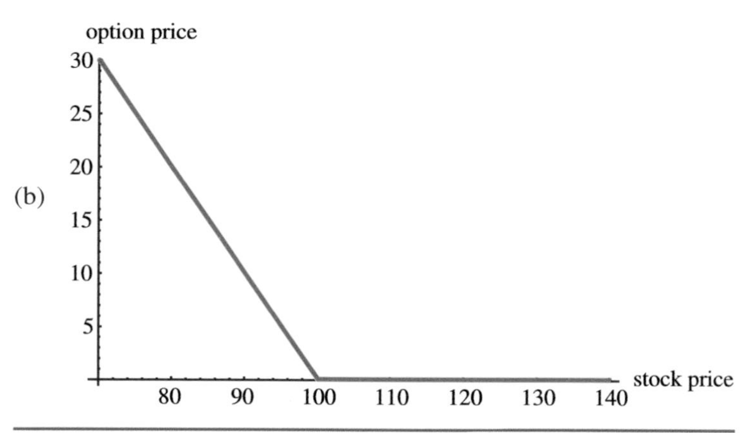


**Figure 1** *(a) The value of a call option on its maturity date. The "strike price" is \$100, which means that the holder of the option can buy the underlying stock for \$100 (no matter what the market price is on that date). If the market price is \$100 + x (where x is positive), then the bearer can make a profit of \$x. Otherwise, the option is worthless. (b) Conversely, a put option has value only if the price of the stock falls* **below** *the strike price. Thus, put options are useful as a hedge against a drop in asset prices.*

**2Q 2008**
US GDP increases by 0.6 percent (annual rate).

**7/12/2008**
IndyMac bank closes, taken over by FDIC. Largest US bank to fail since 1984.

**7/14/2008**
Paulson proposes rescue plan for Freddie Mac and Fannie Mae.

**9/7/2008**
US government takes control of Freddie Mac and Fannie Mae.

**9/15/2008**
Lehman Brothers announces largest bankruptcy filing in US history.

bank account would occur over that time. It also contains a stochastic part ($\sigma dX$) that is described by a Brownian motion, the same kind of random wiggling that describes atoms in a gas. The parameter $\sigma$, called the volatility, describes how big the random fluctuations of the stock price are likely to be.

Under this assumption, Black, Scholes and Merton showed that the value of the option, $V(S,t)$, satisfies a deterministic (not stochastic) differential equation that is reminiscent of the heat flow equation in a gas. At a time $t$ before the maturity date, the sharp corner in the value of a put or a call on the maturity date gets smoothed out. (See Figure 2, next page.) If the option is "out of the money" at time $t$, its value is still greater than zero, because there is still time before the maturity date for random fluctuations to move the stock's price back up over the strike. Likewise, if the option is "in the money," it is still slightly more valuable than it would be on the maturity date, because of the asymmetry between the amount it can gain and the amount it can lose. However, the higher the current stock price, the smaller this gap becomes.

The key point to the Black-Scholes formula was that it was deterministic. The fair market value of an option could be calculated precisely, even if the fluctuations of the underlying security were unpredictable. All you needed to know were a few parameters, such as the volatility, the strike price, and the time to maturity—and you needed to know how to solve a differential equation. It was an open invitation to math and physics degree holders to come to Wall Street and become "quants."

Of course, the Black-Scholes model was not perfect. To begin with, there are many more sophisticated options and "derivatives" than the simple call and put options illustrated here. Also, some of the assumptions underlying the model are debatable. It assumes a perfectly liquid market. The model of volatility (with constant volatility $\sigma$) is too simplistic in many cases. And since the stock market crash of 1987, there has been an asymmetry in market prices that Black-Scholes can't account for. Put options that are "out of the money" tend to have a higher market price than the model predicts, perhaps

**9/15/2008**

Bank of America announces purchase of Merrill Lynch for $50 billion.

Major credit agencies downgrade AIG credit, Federal Reserve Bank of New York fails to broker private deal to bail out AIG.

**9/16/2008**

US Federal Reserve announces $85 billion bailout of AIG, will acquire 79.9 percent stake in company.

**9/19/2008**

SEC announces temporary ban on short selling on 799 US financial stocks. This move is followed by similar bans in the UK, Australia, Taiwan, and the Netherlands.

Treasury Secretary Paulson proposes $700 billion bailout (the Troubled Asset Relief Program, TARP) to enable US government to purchase "toxic assets" from banks.

**9/22/2008**

Goldman Sachs and Morgan Stanley become bank holding companies; all major investment banks in the "shadow banking system" are now either dissolved or regulated.

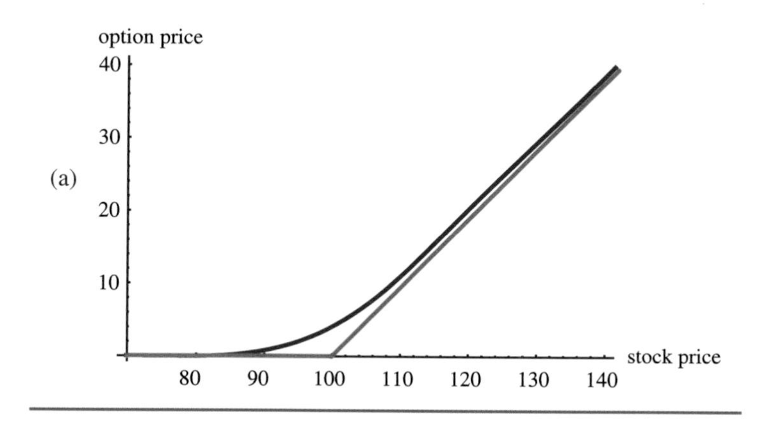


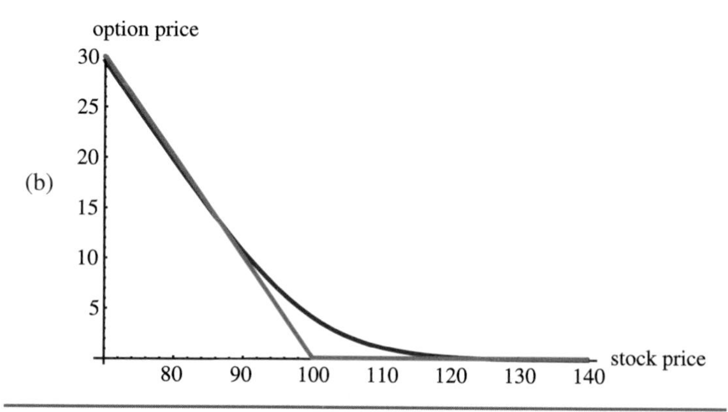


**Figure 2.** *The Black-Scholes formula makes it possible for investors to compute the value of an option before its maturity date, provided that they know the volatility of the underlying stock. (a) The blue curve shows the value of a call option at 0.1 years before the maturity date. (b) The blue curve shows the value of a put option at 0.1 years before the maturity date. As in Figure 1, the red curve shows the value on the maturity date. (Figures courtesy of: "Black-Scholes Option Model" from The Wolfram Demonstrations Project, http://demonstrations.wolfram.com/BlackScholesOptionModel. Contributed by: Michael Kelly (Stuart GSB, Illinois Institute of Technology).)*

**9/26/2008**

Washington Mutual, the sixth-largest bank in US, fails, the largest bank failure in US history. JP Morgan Chase acquires WaMu's assets.

**9/29/2008**

U.S. House of Representatives votes against bailout package. Citigroup buys Wachovia, the sixth-largest lender in US, with assistance from FDIC.

**3Q 2008**

US GDP decreases by 4.0 percent (annual rate).

**10/2/2008**

Revised bailout plan passes Congress, signed by President Bush.

**10/8/2008**

AIG bailout increased by $38 billion.

**10/2008**

Credit crisis spreads to Europe, as several governments guarantee bank deposits and inject currency into their leading banks. Iceland nationalizes the three largest banks in the country.

because investors purchase them as insurance against future crashes (see Figure 3). This has led some people, such as Merton, to add a third term to the differential equation that describes large market swings as a separate phenomenon from day-to-day volatility.

However, none of these limitations meant that Wall Street had to give up on Black-Scholes. Like any scientific model, it simply needed to be revised. "People realized it's a useful framework, but you have to make adjustments," Lo says. "There was a version 2.0, and now we're up to something like version 6.8. It was still an important piece of intuition and a theoretical breakthrough."

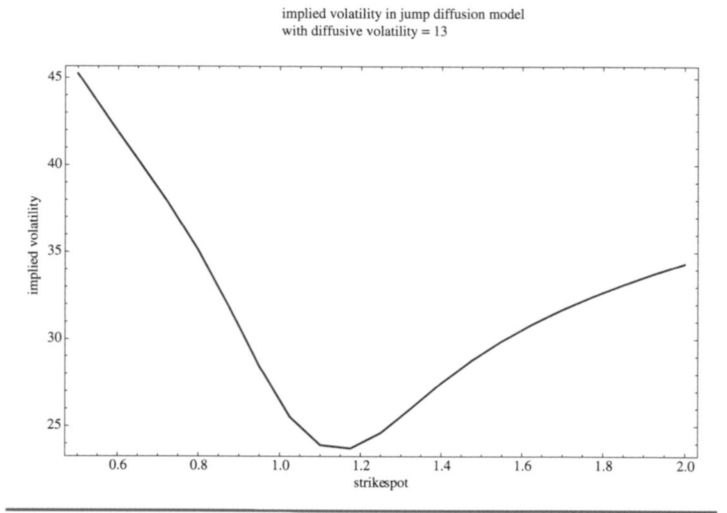


**Figure 3.** *A flaw in the Black-Scholes model is the "volatility smile." Out-of-the-money options tend to be priced by the market as if the volatility of the stock were higher. Thus a graph of the "implied volatility" bends upward or "smiles" at each end. A modification to the Black-Scholes model called jump diffusion takes this behavior into account. (Figure courtesy of: "Black-Scholes Option Model" from The Wolfram Demonstrations Project, http://demonstrations.wolfram.com/BlackScholesOptionModel. Contributed by: Michael Kelly (Stuart GSB, Illinois Institute of Technology).)*

**10/11/2008**
Dow Jones Industrial Average finishes its worst week in history.

**11/9/2008**
AIG bailout package increased to $150 billion.

**12/11/2008**
National Bureau of Economic Research determines that US has been in a recession since December 2007.

**12/19/2008**
President Bush announces $17 billion in loans to General Motors and Chrysler out of the TARP.

**4Q 2008**
US GDP decreases by 6.8 percent (annual rate).

**1/10/2009**
Unemployment rate rises to 7.2 percent.

"There is a very good analogy between Black-Scholes and the Gaussian copula formula. In fact, it's more than an analogy, it's an exact equivalence," Lo says. "The copula was version 1.0 of a very complex set of mathematics. It provides intuition and guidance, but it can't be used at all times in all settings. In version 2.0 we'll see some differences. We still need to get to that version 6.8."

## Subpriming the Pump

From the viewpoint of an average person, the trigger for the financial crisis of 2007 and 2008 had little to do with mathematics. In the middle of the decade, roughly between 2004 and 2006, housing prices increased dramatically. Prior to 2001, the national median home price was about 3 times the median household income; by 2006, that ratio had risen to 4.6 times. In order to pay for those homes, buyers had to take out larger and riskier mortgages, which in ordinary times they might not have qualified for. Such loans, which entail a higher than normal risk of default, are called "subprime."

It is easy to blame home buyers for taking out loans that they couldn't afford. But that would be missing half the picture. Why were the banks so willing to make the loans? The answer is that they had a ready market of investors—both at home and abroad—who were eager to assume the risks, in exchange for a crack at the profits in a booming housing market. The key device that made this possible was a mathematical method for pooling the risk, called "tranching."

The **"Securitization"** figure on page 28 shows a very simplified example of how it works. The bank that made the loans can bundle two of them up into a security called a "collateralized debt obligation" (CDO). The CDO can be sliced into different layers of risk ("tranche" is the French word for "slice"). For example, an investor in the senior tranche will make $1000 if *either* borrower repays his loan, and she will only lose money if *both* borrowers default. In ordinary times, this

| 2/17/2009 | 3/2/2009 | 3/19/2009 | 1Q 2009 | 4/30/2009 | 5/7/2009 |
|---|---|---|---|---|---|
| President Obama signs a $787 billion economic stimulus package. | AIG reports largest single quarter loss by a US company in history ($61.7 billion, 4Q 2008); Fed and Treasury announce restructuring plan for AIG that could allow it to receive $30 billion more from TARP. | FDIC sells IndyMac to OneWest Bank. | US GDP decreases by 4.9 percent (annual rate). | Chrysler files for bankruptcy. | "Stress tests" by Federal Reserve determine that 9 of 19 largest US bank holding companies have adequate capital in case of a worst-case scenario for the 2009-2010 economy. The 10 other firms are required to raise more capital. |

would be very unlikely, and the senior tranche would be a much safer investment than either of the individual mortgages. Meanwhile, the bank can offer a junior tranche to another investor. This one only pays off if *neither* of the borrowers default. It is a riskier bet, but on the other hand it is priced lower, so the investor in the junior tranche can make more money.

If you scale this example up, by pooling together not two but thousands of mortgages, you can create investments with virtually any desired degree of risk and return, from ultra-safe to ultra-speculative (but highly profitable). "It's like taking crude oil and decomposing it into different grades, which can be used for different purposes—plastics, natural gas, gasoline," says Lo. "No one wants crude oil, but they want the products you can make from it."

The tranche mania suddenly made vast quantities of capital available to homebuyers in the 2000's. That was why loans could be approved that would not have been approved before. The capital was not coming from conventional, commercial depository banks. It was coming from what many economists call the "shadow banking system"—investment banks like Lehman Brothers and Merrill Lynch that did not have the same rules and limitations that commercial banks do.

"The subprime crisis took an innovative idea and grew it to ridiculous proportions," Lo says. "It was a pipeline of money from investors to the housing market. It was like discovering offshore oil for the first time." And just like offshore oil drilling, it had risks that were not fully appreciated at the time.

Have you spotted the flaw yet in the **"Securitization"** Figure on page 28? If not, you're not the first person to miss it. The flaw is that we assumed that the two borrowers' risks of default were independent. That was why we said that the probability that both of them would default was 1 percent (10 percent times 10 percent). In the real world, they might not be independent. In fact, they could be completely dependent—whenever one defaults, the other one does too. Maybe they both worked for the same company and lost their jobs at the same time. Maybe they were both counting on refinancing when their house's value went up—and when the housing bubble finally burst, at the beginning of 2007, they couldn't get a loan any more.

| 6/1/2009 | 6/10/2009 | 6/25/2009 | 2Q 2009 | 3Q 2009 | 10/14/2009 |
|---|---|---|---|---|---|
| General Motors files for bankruptcy. | 10 banks, including American Express, Goldman Sachs, Morgan Stanley, and JP Morgan Chase qualify to leave TARP program. They can return $68.3 billion of bailout money. | AIG announces it will repay $25 billion of its bailout money. | US GDP decreases by 0.7 percent (annual rate). | US GDP increases by 1.6 percent (annual rate). | Dow Jones Industrial Average closes above 10,000 for the first time in more than a year. |

Whatever the reason, the homeowners aren't the only ones who are out of luck. The owners of the senior tranche are out of luck, too. Instead of a 1 percent risk of losing their capital, they now have a 10 percent risk—and that's not good, when the money they are investing is somebody's retirement account.

Now the story gets more complicated, and the Gaussian copula formula enters the picture.

### Modelling Dependence

The people who created the shadow banking system—the tranches and the CDOs—were not stupid. Or at least, they didn't think they were stupid. They realized that there could be dependence between the default risks of different securities—but they thought that they had the problem licked.

Generally speaking, any *one* of the assets in a CDO will have a well-understood risk of default. It should be noted that the assets in a CDO do not have to be mortgages; often, for example, they are corporate bonds. A quant can use the borrower's history to create a mathematical model of the probability that he (or it) will default before any time $t$. The model may take the form of a "cumulative distribution function," $F(t) = \text{Prob}\{\text{Borrower defaults before time } t\}$.

Now suppose that the quant wants to know the price or the risk for a CDO consisting of two corporate bonds. He knows $F_1(t)$ and $F_2(t)$. But he needs to package them into a two-dimensional cumulative distribution function, $F(t_1, t_2)$, which expresses the probability that bond 1 will default before time $t_1$ and bond 2 will default before time $t_2$. How can he do that?

Long ago, in 1959, a mathematician named Abe Sklar proved a theorem that doesn't so much solve the problem as say that it can be solved. Suppose $F(t_1, t_2)$ is a cumulative distribution function whose "marginal" distributions are $F_1$ and $F_2$ (the separate default risks of the two companies). Then there is a function $C$, which Sklar called a "copula," that links $F_1$ and $F_2$ to $F$:

$$F(t_1, t_2) = C(F_1(t_1), F_2(t_2)).$$

Sklar's theorem, in essence, says that all of the dependence between the two random variables can be packaged into the

**11/1/2009**

CIT Group files for bankruptcy, making it one of the largest financial institutions to receive TARP funds ($2.3 billion) and still fail.

**11/9/2009**

FDIC announces that 9 of the 10 banks that failed the "stress test" have now raised sufficient capital. Only GMAC has not raised enough.

**12/9/2009**

Bank of America repays TARP funds ($45 billion).

**12/23/2009**

Citibank ($20 billion) and Wells Fargo ($25 billion) repay TARP funds.

All major U.S. banks have now repaid their bailout money; however, many mid-sized and regional banks continued to fail throughout 2009 and 2010 **(see Figure 4).**

**4Q 2009**

US GDP increases by 5.0 percent (annual rate).

function $C$, which is defined on the unit square $[0,1] \times [0,1]$. For example, if $F_1$ and $F_2$ are independent, then

$$C(x_1, x_2) = x_1 x_2.$$

This simply says that if the probability the first lender defaults before time $t_1$ is $x_1$ and the probability that the second lender defaults before time $t_2$ is $x_2$, then the probability of both events is $x_1 x_2$. Another easy example is this: If the two lenders are perfectly dependent, so that one defaults whenever the other does, then $C(x_1, x_2) = \min\{x_1, x_2\}$.

In 2000, a quantitative analyst at JP Morgan Chase named David Li had a revolutionary idea: Why not turn the (at that time little-known) Sklar's theorem on its head? Sklar's theorem goes *from F to C*. But in quantitative finance, $F$ is what we don't know. So let's just pick a "nice" copula $C$ and go *from C to F*, using Sklar's formula.

As every first-year statistics student knows, the "nicest" one-dimensional probability distribution is the Gaussian (or normal) distribution, whose cumulative distribution function is denoted $\Phi(t)$. And every second-year statistics student knows that the "nicest" two-dimensional probability distribution is the bivariate normal distribution with correlation $\rho$, $\Phi_\rho^2(t_1, t_2)$. So what could be more natural than packaging them together into one copula function,

$$C(x_1, x_2) = \Phi_\rho^2[\Phi^{-1}(x_1), \Phi^{-1}(x_2)]?$$

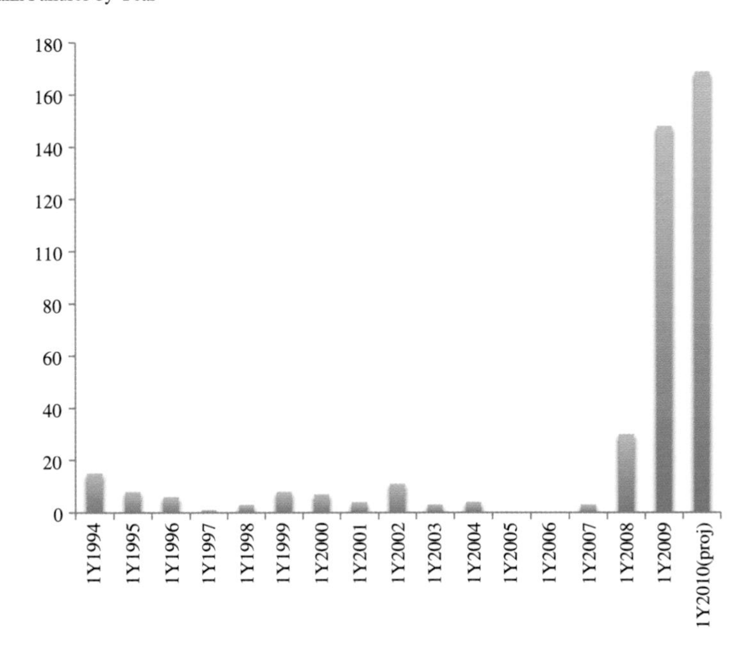


**Figure 4.** *United States bank failures, 1994-2010. Although most of the country's largest banks weathered the crisis in 2008 and 2009, with assistance from the government, smaller banks continued to fail throughout 2010. (Source: Federal Deposit Insurance Corporation. Projection for 2010 based on data through 8/17/2010.)*

**For all these reasons, the Gaussian copula should have been a good starting point, a first stab at the problem of pricing and evaluating the risk of CDO's. . . Instead it became the default model. . . Now it wasn't the formula abdicating responsibility—it was *people*.**

Thus was born "the formula that killed Wall Street," the Gaussian copula formula.

The Gaussian copula has some distinct advantages. It is easy to calculate, because these functions are programmed into every statistical package. And it reduces the dependence relation between the two assets to a single number, the correlation $\rho$. To the quant, it seems as if he is protected. If your underlying assets are highly correlated, you can build that into the copula and estimate the risks accordingly.

The flaws of the Gaussian copula, however, are numerous. First, it abdicates the problem of truly understanding the correlation between assets, replacing it with a single number. Second, it assumes that number is constant over time. There was very little correlation between mortgage defaults in Miami and defaults in Las Vegas as long as housing prices were going up. But when prices started going down—all over the country—the correlations changed. Third, as Taleb points out, the validity of any economic model based on Gaussian distributions is questionable. The bell curve has very skinny tails, which means that it underestimates the likelihood of large deviations from the mean (see Box, **"Heavy Tails and Financial Markets"**, page 42). Finally, not only are the tails too skinny, but the Gaussian copula links them in such a way that "tail events" are nearly uncorrelated (no matter what value $\rho$ takes). By using the Gaussian copula function you are assuming that in "normal" times the default risks are correlated, but in extremely unusual times they are not. That is the exact opposite of reality.

For all these reasons, the Gaussian copula should have been a good starting point, a first stab at the problem of pricing and evaluating the risk of CDO's—Lo's "version 1.0." Instead it became the default model. By 2004, all three credit rating agencies had accepted the Gaussian copula as a way of evaluating the credit-worthiness of a CDO, while previously they had used rules such as diversification (so that, for instance, mortgage-backed securities would have to be combined with some different kind of asset). Now it wasn't the formula abdicating responsibility—it was *people*.

Another lurking time bomb was the way that the parameters in the model—the marginal distributions of the default times, $x_i(t)$, and the correlations $\rho$—were evaluated. As Lo says, "You can't open up the *Wall Street Journal* and find that information." Instead, Li used the market prices of credit default swaps (CDS)—a sort of insurance policy against defaults—as a proxy for default probabilities. The correlations were "calibrated" (although "guessed" might be a more accurate term) from the CDO prices themselves. The assumption was that if the prices of two bonds were historically correlated in a particular way, then the probabilities of default would be correlated in the same way.

Unfortunately, the CDS and CDO markets had only existed for the last decade, when the housing market was going up, defaults were rare, and all of the correlations were low (See Figure 5). So even if the copula formula had been perfect (which it wasn't), the parameters plugged into it were entirely wrong after the housing bubble burst. (Continued on p. 46)

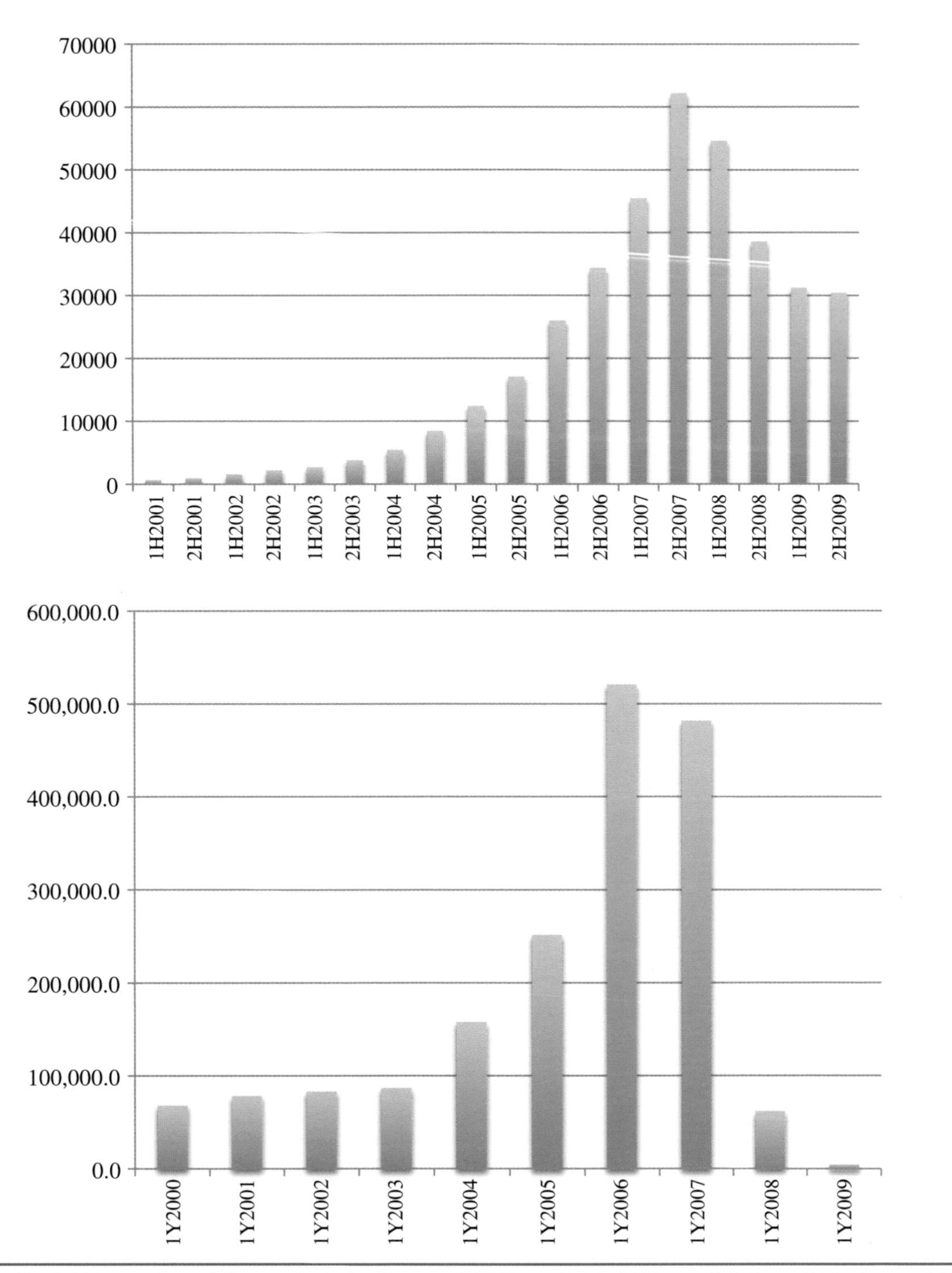


**Figure 5.** *(a) Face value of the credit default swap (CDS) market in billions of dollars. (b) Volume of the collateralized debt obligation (CDO) market in millions of dollars. Both markets were essentially nonexistent before 2000. The CDO market dried up after the financial crisis of 2007. The CDS market has continued, but at a much reduced level. (Sources: (a) International Swaps and Derivatives Association, Inc. (b) Securities Industry and Financial Markets Association.)*

### Heavy Tails and Financial Markets

One common criticism of the Gaussian copula formula (and other applications of the normal, or Gaussian, distribution to finance) is that it allegedly understates the probability of extreme events. The normal distribution is illustrated in blue in the **Heavy Tails 1** figure.

For example, if the fluctuations in a stock's price were normally distributed, with the $x$-axis representing percentage changes in the stock price over the course of a day, then the probability of a more than 4 percent decrease in the price would be the area under the blue curve and to the left of $x = -4$. In this figure, that probability is less then 1 percent. On the other hand, if the fluctuations were described by the "heavy-tailed" distributions shown in black, red, and green, then the probability of a decrease of more than 4 percent in one day would be considerably higher.

The normal distribution is widely used in practice because it has many desirable properties. First, the distribution depends on only two numbers: the mean $\mu$ (the center of the bell-shaped curve) and the standard deviation $\sigma$ (a measure of the "spread" of the probability distribution; for the blue curve in **Heavy Tails 1**, $\sigma = 2$). If we assume that the stock is fairly priced, then the expected change $\mu$ over the course of one day would be 0. Then a financial analyst would only need to know the value of $\sigma$, called the stock's volatility, to estimate the likelihood of any particular price change. This is a computation that risk managers perform literally every day.

Secondly, if the fluctuations over one day are normally distributed, then so are the fluctuations over any time interval. The only thing that changes is the volatility, which scales as the square root of the time. Thus, if the stock's one-day volatility is 2 percent, then its volatility over 100 days would be roughly 20 percent. [Actually, to make this statement precise, you need to measure the fluctuations logarithmically, using the stock's return. If the stock's price at time $t$ is $p(t)$, then its *return* over a time $\Delta t$ is $\ln(p(t + \Delta t)/p(t))$.]

A more general class of distributions, called Levy stable distributions, has the same desirable scaling property. That is, if the stock's fluctuations for one day follow a particular Levy stable distribution, then the fluctuations over any number of days will follow the same distribution, only with a larger spread. The black, red and green curves in Figure **Heavy Tails 1** show Levy stable distributions corresponding to different values of a "shape parameter" called alpha. When $\alpha < 2$, the distributions have heavy tails. Instead of decreasing exponentially as $x$ gets larger, they decrease according to a power law. For all of these distributions, in fact, the standard deviation is infinite. Thus, if a stock's prices are described by a Levy stable distribution, its "volatility" would be essentially meaningless.

Are stock prices better described by a Levy stable distribution than a normal distribution? Do we truly live in a world of heavy tails?

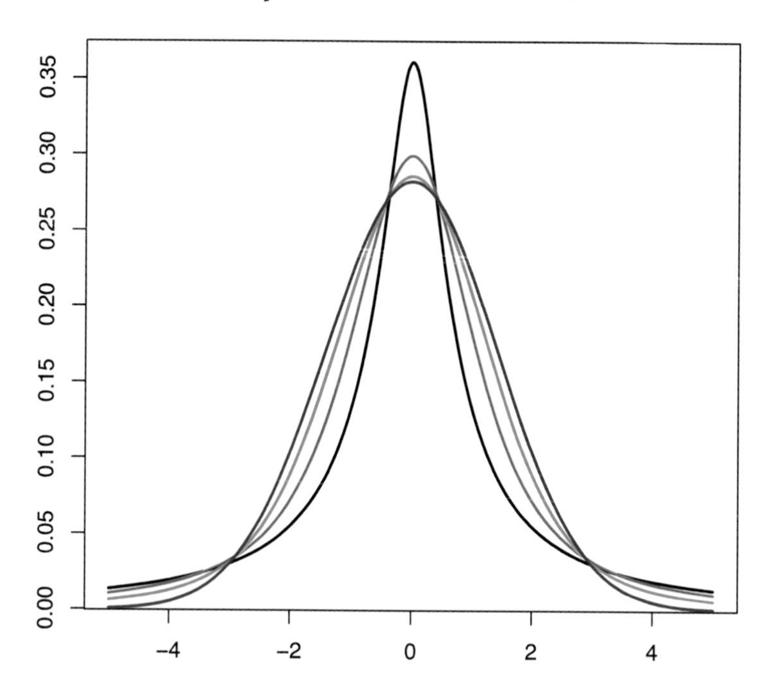


**Heavy Tails 1.** *The Gaussian distribution or "bell-shaped curve" (blue) is often used to model the probability of various changes of asset prices. However, many economists believe that the Gaussian understates the probability of large changes, and that a heavy-tailed distribution (black, red or green) would be a more accurate model. (Figure courtesy of John Nolan; academic2.american.edu/~jpnolan.)*

Figure **Heavy Tails 2**, next page, shows the price and daily returns of Google stock since its initial public offering in 2004. Some extremely large fluctuations can be seen, which suggest that the distribution of returns might indeed have heavy tails. A histogram of the daily returns likewise shows that the best-fitting Levy stable distribution (with $\alpha \approx 1.6$) fits the data better than the normal distribution.

But don't throw out your volatility indexes just yet! The verdict is not so clear. First, it is to be *expected* that some stable distribution would be better than the normal distribution, because the family of Levy stable distributions has more adjustable parameters (four, rather than two). Secondly, where does the improvement lie? The histogram shows a clear improvement in the *central* part of the probability distribution—but it is precisely the *tails* that we are most interested in, because the tails represent extreme swings, the kind that can make or break fortunes.

The final part of **Heavy Tails 2** shows a version of the cumulative distribution function (the integral of the probability distribution function) that has been graphically transformed to emphasize the behavior of the tails. In this graph, a power law should have tails that approach a straight line on either end, while the normal distribution has tails that curve parabolically. The data from Google stock prices do not follow either curve very convincingly. The normal distribution (again in red) is clearly inaccurate; the data lie well below the red curve on the right, and well above it on the left. On the other hand, the data also miss the blue Levy stable distribution by a wide margin, and it is hard to argue that they are following any kind of linear trend at the left and right ends.

All in all, it would seem that mathematics has not yet succeeded in describing the size and frequency of the stock market's largest (and most important!) swings.

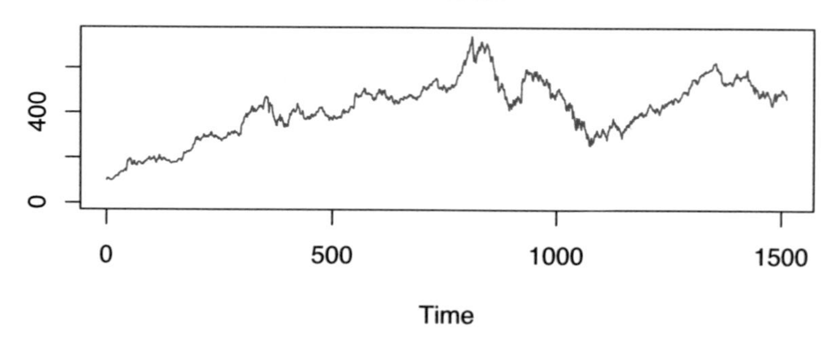


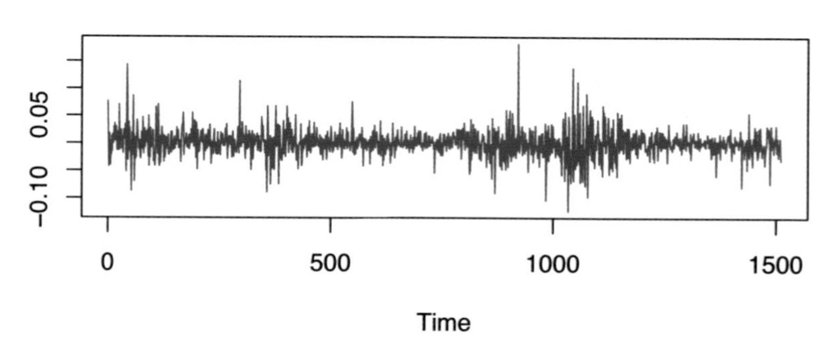


**Heavy Tails 2a.** *Empirical evidence for the "heavy tails" hypothesis is inconclusive. Google stock prices (top) and returns (bottom) since the stock's initial public offering in 2004. A "return" is roughly a daily percentage change. (Figure courtesy of John Nolan; academic2.american.edu/~jpnolan.)*

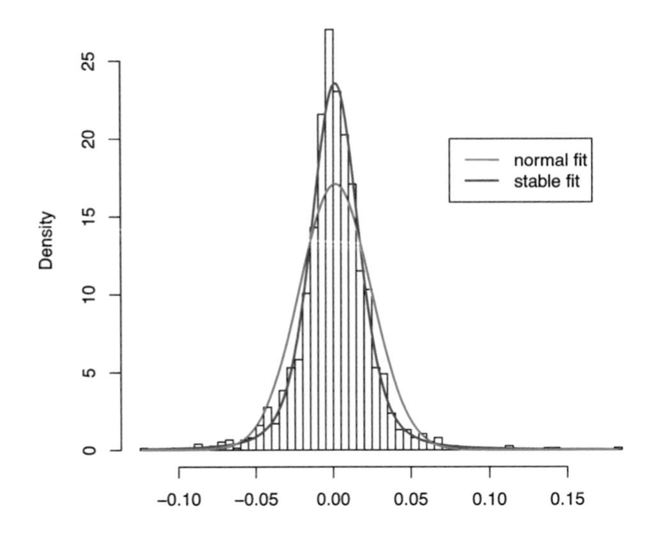


**Heavy Tails 2b.** *A histogram of Google's daily returns fits a heavy-tailed distribution (blue) better than the normal distribution (red).*

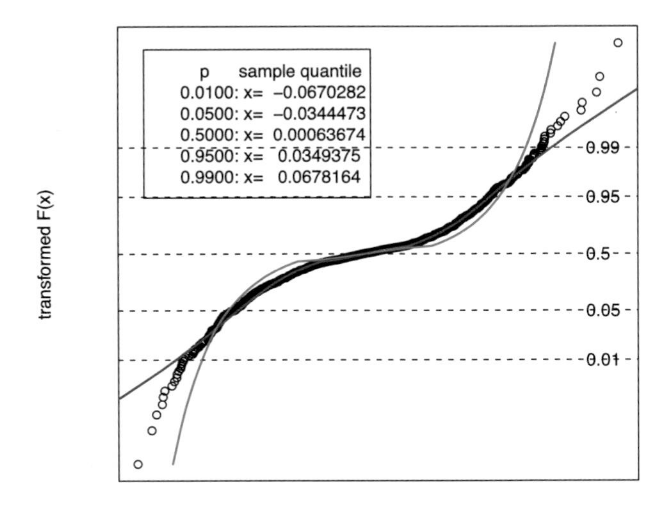


**Heavy Tails 2c.** *However, most of the improvement lies in the center of the distribution. When the distribution is transformed to emphasize the tails, the Google data do not lie very close either to the normal or the heavy-tailed distribution. (Figure courtesy of John Nolan; academic2.american.edu/~jpnolan.)*

**In a perverse way, the financial crisis might be good for mathematical models of credit risk, because it will finally allow more sophisticated models to get a hearing.**

(Continued from p. 40) The credit derivatives market is, by the way, how AIG got sucked into the credit crisis of 2007 and 2008. AIG was a huge seller of credit default swaps, insuring thousands of investors against home loan defaults. "AIG essentially took the other side, a one-way, long-term bet on the U.S. mortgage market," wrote the Congressional Oversight Panel in its June 2010 report on the AIG bailout. Even though the Financial Products division was only a small part of AIG (8 percent by revenue in 2006), when they lost the bet, it was enough to bring down the remaining 92 percent of the company.

How could AIG have wagered the whole company on a bet like this? The congressional report doesn't mince words, and it doesn't blame the Gaussian copula. It says that AIG was "brought down by the company's insatiable appetite for risk and blindness to its own liabilities." In retrospect, examples of this blindness are easy to find. The head of AIG Financial Products said in mid-2007, "It is hard for us, without being flippant, to even see a scenario within any kind of realm or reason that would see us losing $1 in any of these transactions." This comment came at a time when AIG's creditors were already getting nervous and starting to demand collateral.

And AIG wasn't the only offender. To demonize the Gaussian copula formula is to overlook the psychology of the time. Yes, the copula made it possible for quants and their managers to think they had controlled the risk, when in fact they hadn't. But they were living in an era when business was booming, and no one wanted to hear any bad news. "Everybody did so well for so many years," says Lo. "Greenspan was lauded as a hero because he let the markets do their work. It was hard for anyone to spoil the party." Academics pointed out the flaws and the hidden assumptions in the copula model, and suggested alternatives. "More complex models were available, but they weren't widely adopted," says Robert Jarrow, a mathematical economist at Cornell University. "The attitude was, if things aren't broken, why fix them?"

## Sifting Through the Wreckage

In a perverse way, the financial crisis might be good for mathematical models of credit risk, because it will finally allow more sophisticated models to get a hearing. Jarrow, for example, is a proponent of "reduced form" models that take into account only the information known to the market. For example, one approach to modeling a company's default risks has been to assume (in Black-Scholes style) that the company's assets follow a Brownian motion. If you know the company's assets, you can estimate how close it is getting to default. But the market does not know this information; defaults and bankruptcies nearly always come (at least to some extent) as a surprise. Instead, a reduced form model treats default as a stochastic process, with a certain probability of occurring at any time. This leads to a different pricing structure for derivatives based on that probability, such as credit default swaps.

The thorny issue of correlations is also getting a closer look—and already was before the crisis hit. One approach to modeling default correlations, presented in a 2009 paper by Darrell Duffie of Stanford University and Andreas Eckner of Bank of America, assumes the existence of a variable called "frailty," a composite of variables unknown to the market, which causes one firm to become more frail when a second firm defaults. Although the nature of the frailty cannot be known for sure, its size can be estimated from historical data. Duffie and Eckner conclude, for example, that the frailty effect increased the default rate by a factor of three during the dot-com bust of 2001. Though you can't look up frailty in the *Wall Street Journal*, you can look for unexplained deviations of the market from your model results. Instead of sweeping them under the rug, you can interpret them as a warning that the frailty factor is getting larger.

An alternative model of correlations is "default contagion," which treats defaults as a communicable disease. In this case one firm's collapse default directly affects another firm's viability, rather than indirectly affecting it through a shared frailty variable. A very simple contagion model was developed by Mark Davis and Violet Lo of Mitsubishi in 1999, and another one by Jarrow and his graduate student Fan Yu in 2001. At this point, the reduced form contagion models are not compatible with the frailty model (which employs variables that are by definition not observable by the market). However, the diversity of models is not necessarily a bad thing. Wall Street has already experimented with the "one size fits all" approach, and it didn't work.

Because the fixes to the simplistic copula model will necessarily be mathematical, Andrew Lo says, "This is a fantastic time for quants." The masters program in finance at MIT last year had 950 applicants for 60 spots. The graduating class had a 100 percent job placement rate, "in spite of the financial crisis—and, in my view, because of the financial crisis," Lo says.

"It's part of my broader perspective that crises are often the consequences of our technology getting ahead of our ability to deploy it responsibly," Lo says. "In any human endeavor, we learn from our mistakes, like bridge collapses or oil spills. Technology improves because of those mistakes, and the financial industry is no exception. The only difference here was that the accident grew to such proportions that it had a tremendous effect on the global economy. Accidents don't usually affect hundreds of millions of people."

**The only difference here was that the accident grew to such proportions that it had a tremendous effect on the global economy. Accidents don't usually affect hundreds of millions of people.**

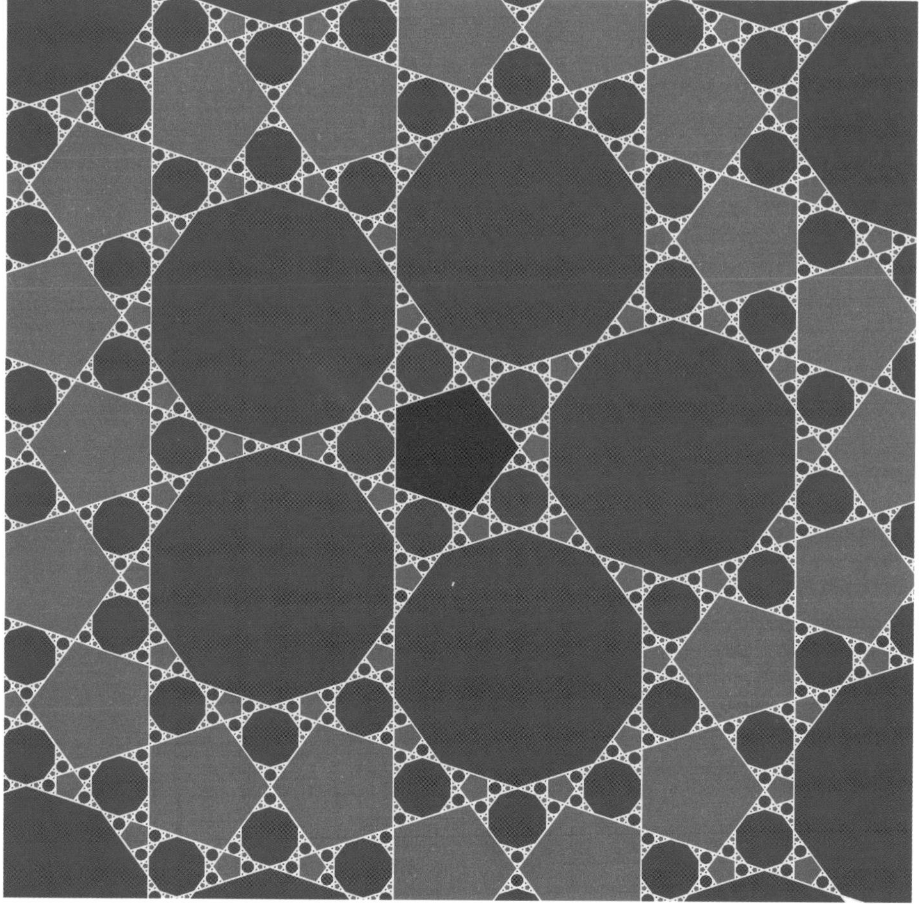

***Orbits of an Outer Billiards Table.*** *Outer billiard orbits for a parallelogram (top) and a regular pentagon (bottom). In each case, all orbits are bounded. (Figures courtesy of Rich Schwartz.)*

# The Ultimate Billiard Shot

RICH SCHWARTZ READILY ADMITS that he isn't much of a billiards player—regular billiards, that is. But in a mathematical variation of the game called "outer billiards," which is played on a computer rather than a smoky saloon, he is the consummate trick shot artist.

Outer billiards is, as the name implies, sort of an inside-out version of regular billiards. The billiard table is infinitely large (which makes it rather difficult to fit into a billiard parlor), and it has a hole in the center. From a starting point outside the region, you shoot the ball so that it just grazes the hole, and continues an equal distance past it (see Figure 1). Then it stops, turns, and repeats the process (each time just grazing the outside of the hole). Unlike a real billiard ball, it experiences no friction and it is free to keep going forever.

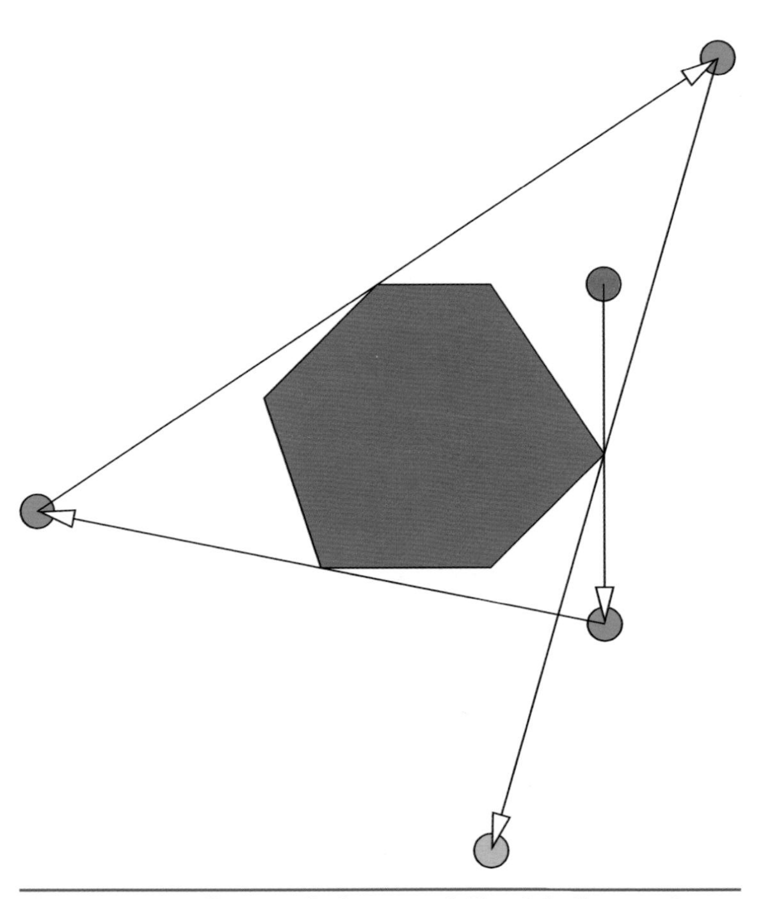

**Figure 1.** *Four "bounces" of an outer billiards ball. At each step, the ball grazes the edge of the "table" (purple) and goes an equal distance past it. (Figure courtesy of Rich Schwartz.)*

**Rich Schwartz.** *(Photo courtesy of Rich Schwartz.)*

The game of outer billiards was first proposed in 1959 by Bernhard Neumann, in a lecture with the unlikely title of "Sharing Ham and Eggs." Ever since then, the biggest question has been: Does the table actually need to be infinitely large? Is there any way a ball that starts near the central region can spiral out to infinity? If the hole is uniformly convex—in other words, its boundary is a smooth curve with no inward dimples or even any flat spots—the answer is no.

However, when some or all of the sides of the hole are straight line segments, the answer depends very sensitively on the shape of the hole. In 2007, Schwartz, a mathematician at Brown University, proved for the first time that for certain shapes, it is possible to play the ultimate trick shot: An outer billiards shot that cannot be contained in any bounded region. In particular, this is true if the hole has the shape of a kite whose long arm is an irrational multiple of its short arm. (See Figure 2; and Figure 3, page 52.)

One characteristic of Schwartz's "trick shots" is that they are extremely sensitive to the starting point. If you move the starting location of the billiard ball by even the tiniest amount, its trajectory may become bounded. By contrast, Dmitry Dolgopyat of the University of Maryland and Bassam Fayad of the University of Paris recently found a billiard table that gives you more room for error. Dolgopyat and Fayad's table has a semicircular hole. On this table, they found a small patch with the property that any billiard ball that is launched from a point within the patch will escape to infinity.

The "game" of outer billiards may seem a little bit contrived because it is hard to imagine how to make a physical billiard ball, all by itself, find a perfect grazing trajectory infinitely many times. However, there are some grounds for considering it a natural counterpart of regular billiards. First, outer billiards on a sphere is mathematically dual to ordinary billiards on a sphere (with the duality map replacing points by lines and vice versa). The idea of going past a vertex by an equal distance is dual to the notion of bouncing off a side at an equal angle, which is what an ordinary billiard ball does. Also, outer billiards is mathematically equivalent to a version of paddleball, which is physically realizable without special equipment like infinite or spherical billiard tables. A ball is attached to a fixed wall by a spring, and it bounces against a paddle that oscillates in a regularly repeating (but not sinusoidal) fashion. The question is whether the ball's bounces can grow larger and larger, *ad infinitum*. The answer, again, is yes.

Outer billiards are also of interest to mathematicians as a "toy model" of planetary motion. The trick is to plot the motion of the billiard ball after every *two* iterations of the billiard map. Suppose the ball starts very far away from the hole, which only looks like a tiny speck in the distance. The first billiard map will take it an equal distance to the opposite side of the speck. The second billiard map will take it back almost to its original position, displaced by an amount roughly equal to twice the width of the (tiny, far away) speck. If you ignore the fact that the ball has made a huge detour, its *net* displacement is very small. When you plot a large number of these small displacements, the ball's trajectory looks like a discrete version of a planetary orbit. If the central shape is an ellipse, then the billiard ball will move in an ellipse (Kepler's first law of planetary motion). The billiard ball's orbit will sweep out equal areas in equal times (Kepler's second law). In general, the billiard map will transform any patch on the billiard table to another patch of equal area—indeed, it is a symplectic mapping (see "A Brave New Symplectic World," page 14).

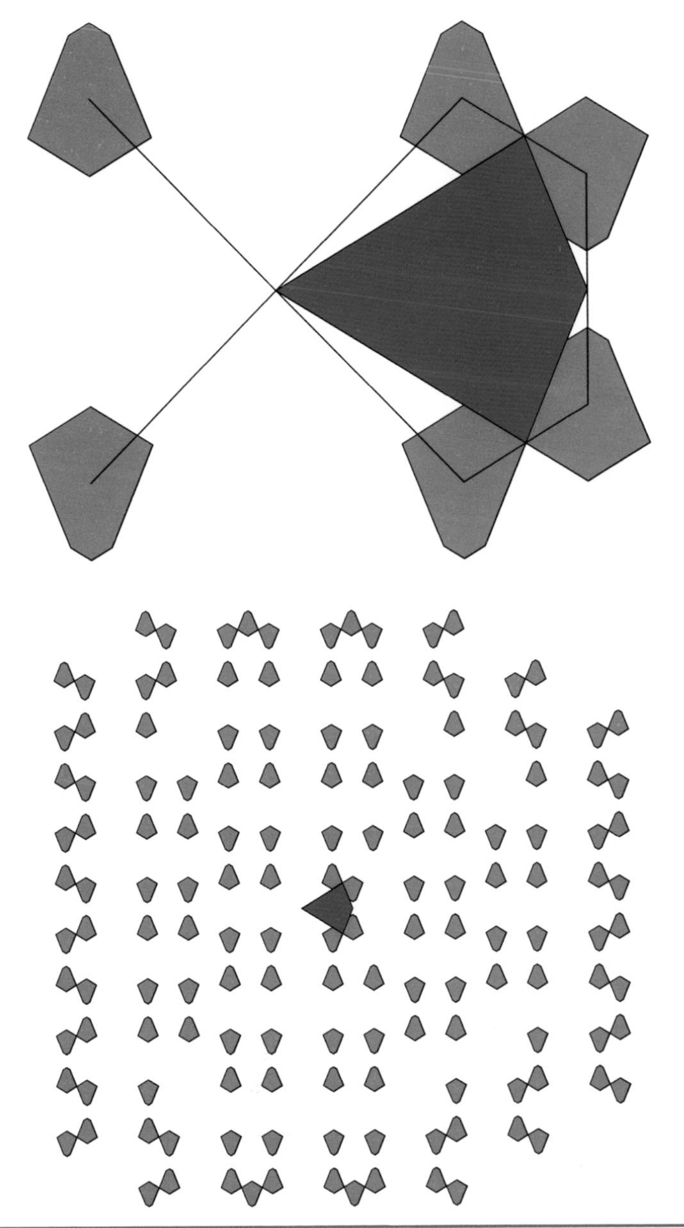

**Figure 2.** *Outer billiard orbits for a rational kite (blue). (Top) Each point in a red region goes to the next red region under the outer billiard map. (Bottom) After finitely many steps, the polygons return to their original place, and the orbits close up. This figure shows the entire orbit of the red polygon in (a). If the outer billiard ball starts at any point in any red polygon, after 150 bounces it will return to its starting point. (Figures courtesy of Rich Schwartz.)*

Drawn by these analogies, Jürgen Moser undertook the first rigorous study of outer billiards in the early 1970s. He proved the boundedness of outer billiard orbits if the central hole is uniformly convex. Moser's proof used a technique,

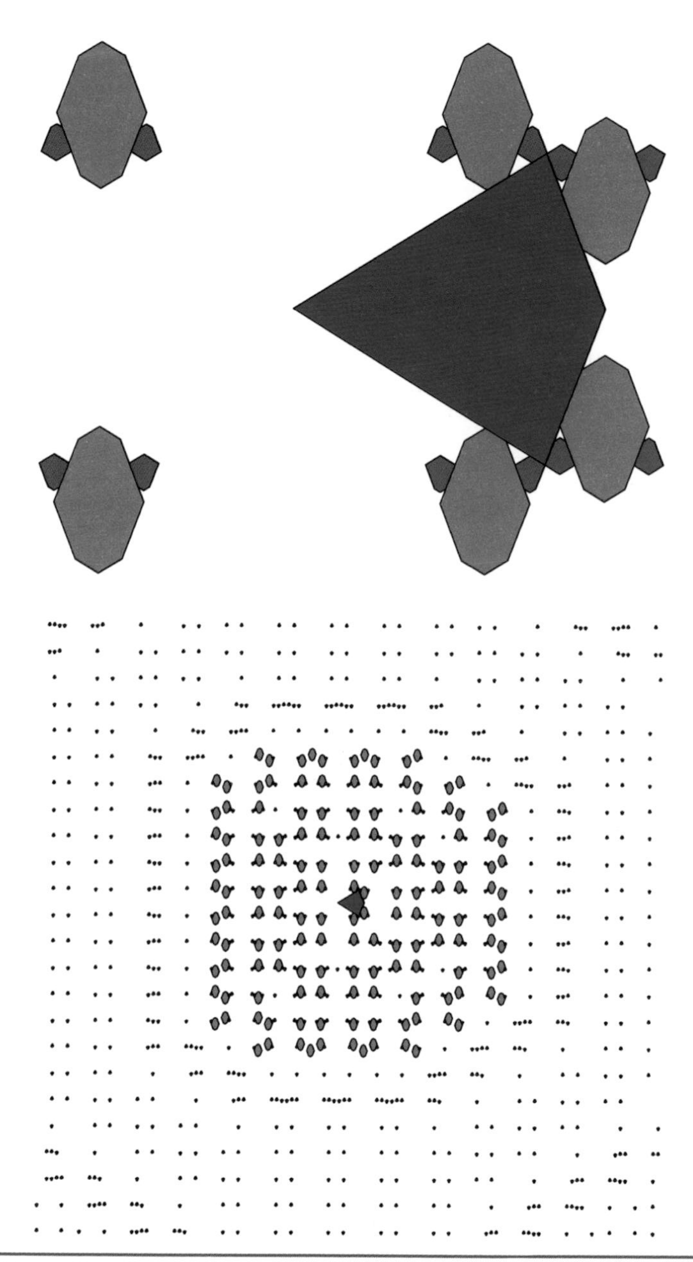

**Figure 3.** *As the blue table gets closer in shape to the Penrose kite (but still rational), more and more complex orbits emerge. (Top) Two basic regions (red and green), with part of their orbits. (Bottom) The complete red orbit is shown, together with only a small part of the green orbit. Because the blue table is a rational kite, no orbits exist that wander out to infinity, but it is reasonable to conjecture that such "wandering" orbits might exist if the kite is irrational. Rich Schwartz has now proved this conjecture to be correct. (Figures courtesy of Rich Schwartz.)*

called KAM or Kolmogorov-Arnold-Moser theory, which can likewise be used to prove that planetary orbits are stable (within certain bounds, which are unfortunately too restrictive to apply to the real solar system).

However, most billiard games are not played on smooth, uniformly rounded tables. The opposite extreme would be polygonal tables. In this case, the billiard map is not continuous, so calculus-based methods like KAM theory will not apply. In fact, the billiard map is not even defined at some points. Remember that the billiard ball is supposed to go to a point of tangency and then continue an equal distance past it. What is it supposed to do if its path is tangent to a whole side of the polygon? Schwartz and others simply make it illegal to place the ball in line with one of the sides. (Also, any starting point that would eventually cause it to reach one of these forbidden points must also be excluded.)

With this modification in the rules, certain results are easy to obtain. If the billiard table is a triangle or parallelogram, then all outer billiard orbits are *periodic*: they always come back to where they started. (See Figure, **Orbits of an Outer Billiard Table**, page 48 (top).) In fact, in both of these cases the shape tiles the plane. Each tile in this figure is mapped by the billiard map to another tile of the same color. So the paths are not only bounded, but they are highly predictable.

In the early 1990s, Sergei Tabachnikov used a computer to study other shapes. For example, he considered the case of a regular pentagon, the first regular polygon that does not tile the plane. In this case, the trajectory of the billiard is not necessarily periodic, but it will still stay within a finite distance of the central hole. In fact, it traces out a beautiful fractal set, as seen in the Figure, **Orbits of an Outer Billiard Table**, page 48 (bottom). In general, three different groups of mathematicians independently showed that for a class of polygons called *quasirational* (including polygons with rational-number coordinates and regular polygons with any number of sides), all orbits are bounded.

Schwartz first heard about the outer billiards problem while he was visiting Tabachnikov at Pennsylvania State University, and he started working on it seriously in 2006. His first step was to write a computer program, which he called BilliardKing, to study outer billiard orbits. To a great extent, the computer remained his collaborator throughout his project. Although the computer could not prove theorems, it suggested patterns and ideas that Schwartz would never have thought of on his own. "I felt like the Lawnmower Man," Schwartz says, referring to a 1992 movie where an unsuspecting man's mental abilities are artificially enhanced. "It made me smarter, a better researcher in a specific way. I haven't done anything like that since."

"It's a new way to do mathematics, very experiment oriented," says Tabachnikov. "It's more like the natural sciences."

Schwartz started experimenting with kite-shaped quadrilaterals—the simplest shape not covered by any of the previous theorems. He started with kites with rational coordinates because the computer can do rational arithmetic without any roundoff error. When he experimented with kites whose legs formed a ratio of Fibonacci numbers, he started to notice some "interesting and vaguely predictable patterns," he says.

**Although the computer could not prove theorems, it suggested patterns and ideas that Schwartz would never have thought of on his own. . . "It made me smarter, a better researcher in a specific way. I haven't done anything like that since."**

This led him to one of the most celebrated shapes in recent mathematical history: the Penrose kite. This is a quadrilateral whose long diagonal is $(1+\sqrt{5})/2$ or about $1.618\ldots$ times its short diagonal. The number $\varphi = 1.618\ldots$, called the golden ratio, itself has a very long history in mathematics; it is the limit of the ratios of consecutive Fibonacci numbers $(8/5, 13/8, 21/13$, etc.). Along with a second shape called the Penrose dart, the kite can be used to construct infinite but non-repeating tilings of the plane. Examples of such "aperiodic tilings" had been known before Roger Penrose, a British physicist, discovered the kite and the dart in the 1970s. However, it is fair to say that Penrose's creation had a beauty and simplicity that was lacking in earlier examples. Now the same shape seemed to be related to a brand new kind of non-periodicity.

"I still remember the magic moment on the bike path when I did the calculation that my kite was really the same as the Penrose kite," Schwartz says. "At the time I still hadn't proved anything, but this little realization made me feel as if a voice from above was telling me to continue working."

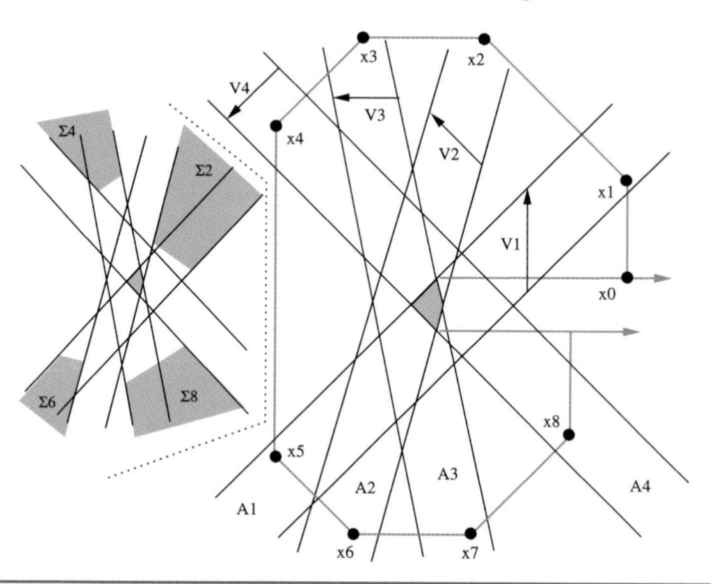


**Figure 4.** *Doubly-iterated outer billiard map far from the Penrose kite. After every second bounce, the ball will return nearly to where it was before, but will take one "step" counterclockwise around the table. After enough steps, it will arrive at one of the four strips shown, each of twice the width of the Penrose kite. At that point, the ball's trajectory turns and it will proceed in a straight line to the next strip, where it will turn again. Close to the kite, where the strips overlap, the dynamics of the ball are much more complicated. (Figures courtesy of Rich Schwartz.)*

Schwartz placed the vertices of the kite at the points $(-1,0)$, $(0,1)$, $(0,-1)$, and $(A,0)$. Here, $A$ represents the cube of the reciprocal of the golden ratio, $A = (1/\varphi)^3$. In a key simplification, he decided to look only at orbits that started at a point on the lines $y = 1$ or $y = -1$, and to the right of the kite. A billiard ball that starts at such a point will always go to a point with an odd $y$-coordinate. Eventually, it will return to either the line $y = 1$ or $y = -1$. This is very easy to see if the billiard ball starts very far away from the kite. (See Figure

4.) Again plotting only the even iterates of the billiard map, we would see the billiard ball gradually moving around a large octagon. At each step it moves either zero, two, or four units in the $y$-direction. Each time it enters one of the strips shown in Figure 4, it makes a one-eighth turn and starts moving along the next side of the octagon. After eight turns, it has made one complete circuit around the kite, and it must come back either to the line $y = 1$ or $y = -1$; it can't jump over both of them because each step is too small.

This argument only works if the starting point is far enough away from the kite; closer to the kite the ball can take bigger vertical steps and miss the lines $y = \pm 1$. However, near the kite the trajectories can be analyzed by brute force, because there are only finitely many distinct types of trajectory. "You can plot a million points and see that it always works," Schwartz says.

Thus, Schwartz proved the existence of the "first return map," which maps any point on the lines $y = \pm 1$ to the next encounter of the billiard ball with those two lines. Next, Schwartz noticed that if the billiard ball starts at a point with coordinates $(2m + 2nA + \epsilon, \pm 1)$, where $m$ and $n$ are integers, then it will always come back to a point of the same form. The numbers $m$ and $n$ serve as a sort of "address" for each point on the billiard ball's trajectory. The number $\epsilon$ is an arbitrary fudge factor that he calls the offset. In his paper, he chose $\epsilon = (1 - A)/2$ because it gave him especially nice results.

This gave Schwartz an idea. Instead of looking at the trajectory in physical space, he looked the return map in what one might call "address space." The points $(m, n)$ in address space, he discovered, trace out a jagged path like the one in Figure 5a (next page), which he calls the "Arithmetic Graph."

Seeing this picture, Schwartz recalls, was his second "aha!" moment. "This figure has the whole paper in a nutshell. The first moment I saw it, I knew that I would eventually solve the conjecture," he says.

The open circle marks the one of the two best starting points for Schwartz's "trick shot": the point $((1 - A)/2, -1)$ in physical space, which corresponds to $(0, 0)$ in address space. Distance from the kite in physical space corresponds roughly to distance above the lower boundary line in address space. Starting at the circled point, the billiard ball's trajectory in address space (bold) appears to meander infinitely far up and to the right. Thus the ball also travels infinitely far from the kite in physical space.

Incidentally, the backwards trajectory (also in bold) of the billiard ball is equally interesting, perhaps even more so. It wanders up and to the left, traveling farther and farther from the kite but always coming close to it again. Of course, it never comes back to exactly the same position. Such a path is called "erratic," and it was a behavior that also had never been observed in outer billiards.

However, as generations of mathematics students have learned, a picture is not a proof. The challenge for Schwartz was to prove these statements correct. Schwartz kept finding more and more clues in the pictures. First, he noticed the scaling properties of the Arithmetic Graph, in which Fibonacci numbers run rampant. The graph has a little "toe" that ends at the point $(0, 0)$. It has a second toe at $(-5, 1)$. These two

**However, as generations of mathematics students have learned, a picture is not a proof. The challenge for Schwartz was to prove these statements correct.**

toes form opposite sides of a "foot." Then there is a sort of deformed foot that doesn't quite make it down to the baseline, and a second foot that does. The first toe on this foot has coordinates $(-21, 5)$. Zooming out, as in Figure 5b, you can

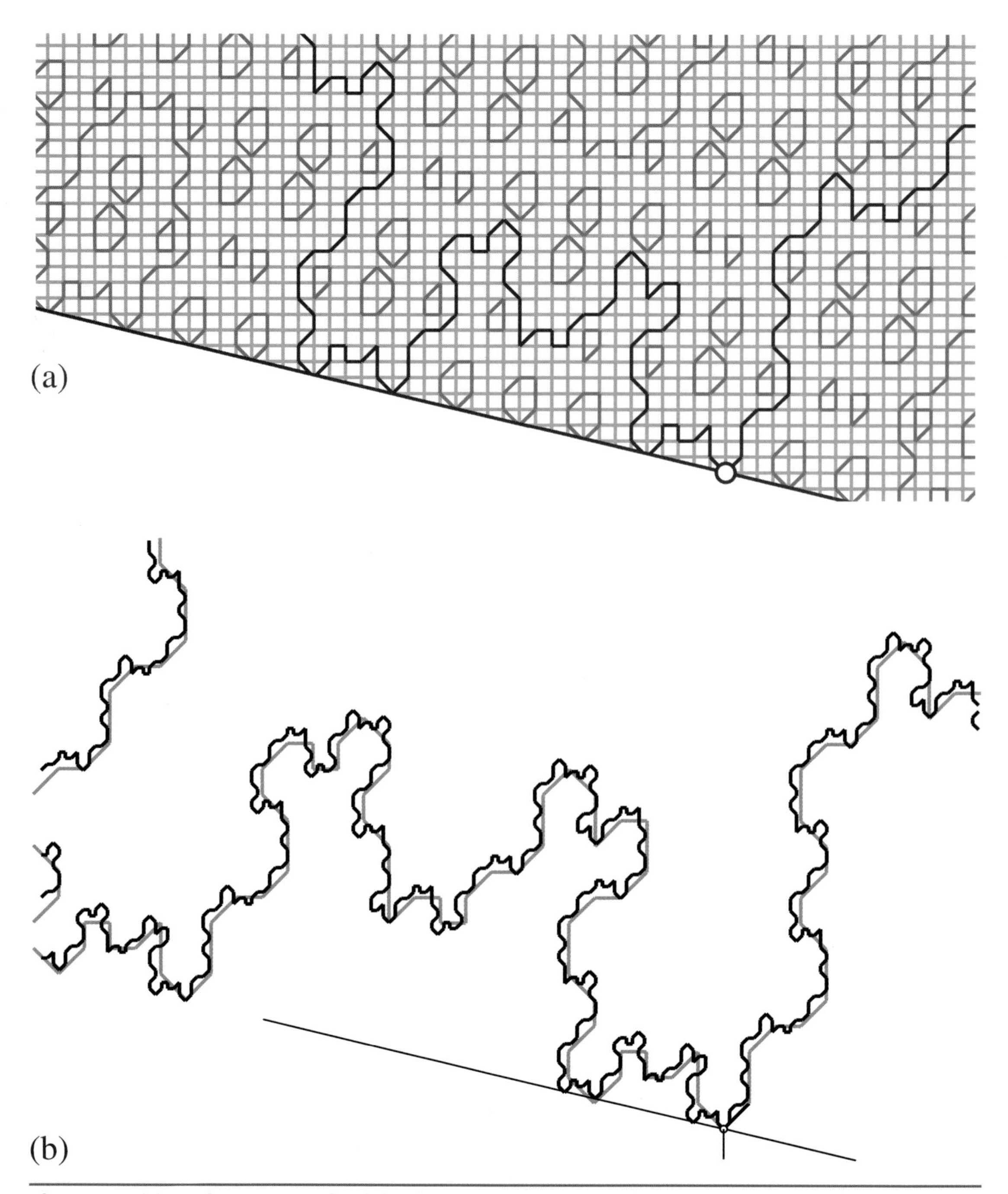


**Figure 5.** *(a) Arithmetic Graph of the outer billiard ball for the Penrose kite. A discontinuous coordinate transformation takes orbits of the billiard ball to an integer lattice. This presentation of the orbit makes the "wandering" orbit (black) easy to find. More importantly, it reveals a fractal structure, shown in (b). When the Arithmetic Graph (top) is magnified (gray), the scaled-up version remains extremely close to the original version (black), a property called self-similarity. (Figures courtesy of Rich Schwartz.)*

see a pedestal, which consists of two feet, each one with two toes. Where does this second pedestal begin? At $(-89, 21)$. The numbers $1, 5, 21, 89, \ldots$ are every third Fibonacci number:

$$1, \mathbf{1}, 2, 3, \mathbf{5}, 8, 13, \mathbf{21}, 34, 55, \mathbf{89}, \ldots$$

Thus, each time you blow up the picture by a factor of $1/A$ (or the cube of the golden ratio), *you get a picture that looks essentially the same.* A magnified toe looks like a foot. A magnified foot looks like a pedestal. A magnified pedestal looks like a super-pedestal. And so on. It is reminiscent of Augustus De Morgan's 19$^{th}$ century parody of a poem by Jonathan Swift:

"Great fleas have little fleas upon their backs to bite 'em,<br>
And little fleas have lesser fleas, and so *ad infinitum.*<br>
And the great fleas themselves, in turn,<br>
have greater fleas to go on,<br>
While these again have greater still,<br>
and greater still, and so on."

This property, called self-similarity, means that the trajectory starting at $(0, 0)$ has to move infinitely far away from the baseline in address space. The reason is that once it moves, say, 10 units away, by self-similarity it must move $10\varphi^3$ units away; then it must move $10\varphi^6$ units away; and so on. The actual proof was a bit more subtle because the self-similarity is not precise. As seen in Figure 5b, the magnified curve (in gray) is not a precise match for the Arithmetic Graph (in black). However, Schwartz showed that the grey and black curves never stray more than 4 units away from each other. This is close enough to allow the argument above to work. Thus, the billiard must move farther and farther away from the kite in physical space.

Self-similarity is notoriously easy to "see" in a picture, but notoriously hard to prove rigorously. The proof took Schwartz about six months, and required more computer-guided insights. With a degree of forthrightness seldom seen in a research paper, he wrote: "The directness of this discussion is misleading. We arrived at the ideas here only after having exhausted every half-baked and useless scheme we could think up."

Next, Schwartz asked, what about other kites? Kites with a rational aspect ratio have only bounded orbits, but he suspected that any kite with an irrational aspect ratio would have unbounded orbits. In other words, there was nothing particularly unique about the golden ratio. Remarkably, he succeeded in proving this, but it took a completely different approach. The approximate self-similarity manifested in the case of the golden ratio does not hold true in general. The golden ratio has a repeating continued fraction, and this fact leads directly to the very nice scaling properties of its Arithmetic Graph. In general, if a number $r$ is a quadratic irrational (i.e., it can be written using nothing but ordinary arithmetic operations and *one* square root sign), then its continued fraction expansion will have a repeating pattern. In this case, the Arithmetic Graph of the kite with aspect ratio $r$ is probably self-similar,

**Self-similarity is notoriously easy to "see" in a picture, but notoriously hard to prove rigorously. The proof took Schwartz about six months, and required more computer-guided insights.**

although Schwartz has not yet proved this. However, an irrational number that *cannot* be written with a single square root (for example, $\pi$, $e$, or $\sqrt[3]{2}$) does not have a repeating continued fraction. If the aspect ratio of a kite is one of these numbers, then its Arithmetic Graph does not appear to be self-similar.

Nevertheless, by experimenting with his BilliardKing program, Schwartz discovered some amazing properties that replaced self-similarity. He considered the Arithmetic Graphs for two kites with aspect ratios $(p_1/q_1)$ and $(p_2/q_2)$, where the numerators and denominators are odd numbers and the two fractions are "unusually close." That is, the numerator of the difference, $(p_1/q_1) - (p_2/q_2)$ is 2, which is the smallest it can be without being zero. An example of "unusually close" fractions would be 7/11 and 31/49 because $7/11 - 31/49 = 2/539$. A look at the Arithmetic Graphs for these two kites (see Figure 6) reveals a surprising fact: the Arithmetic Graph for 31/49 contains two identical copies of the Arithmetic Graph for 7/11!

This is no accident. Still assuming that $(p_1/q_1)$ and $(p_2/q_2)$ are unusually close, Schwartz proved what he called his Copy Theorem. If $q_2$ is the larger of the two denominators, the Arithmetic Graph for $p_2/q_2$ contains roughly $q_2/2q_1$ exact copies of the Arithmetic Graph for $p_1/q_1$, followed by a larger "bump" that travels at least $q_2/q_1$ times farther from the critical line than the Arithmetic Graph for $p_1/q_1$. Figure 6 shows the Arithmetic Graphs for the two kites with $p_1/q_1 = 7/11$ and $p_2/q_2 = 31/49$. Because 49/2(11) is a little bigger than 2, Schwartz's Copy Theorem says that the Arithmetic Graph for 31/49 should contain 2 exact copies of the Arithmetic Graph for 7/11, followed by a bump that is at least 4 times higher than the Arithmetic Graph for 7/11, and that is exactly what the figure shows.

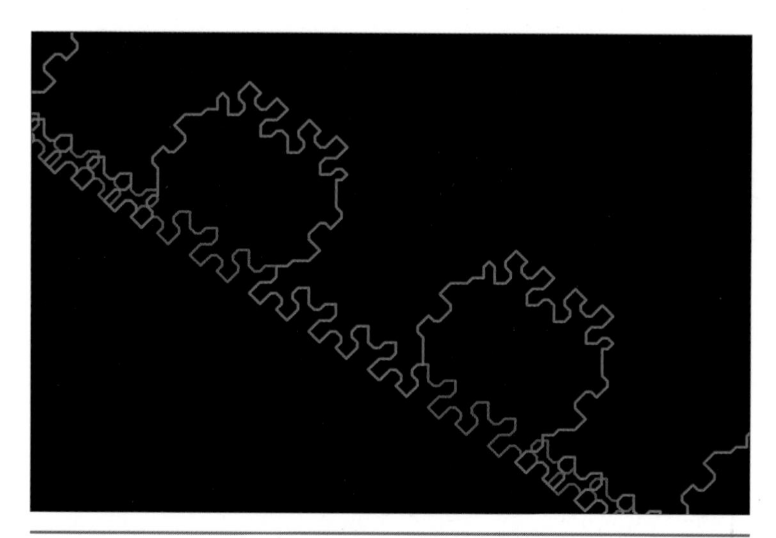

**Figure 6.** *Arithmetic Graphs for kites with aspect ratios 7/11 (blue) and 31/49 (red). Note the large region of overlap, and note that the "bumps" of the red graph are more than twice the size of the bumps of the blue graph. As aspect ratios get closer and closer to an irrational number, the length of the overlap region will increase and the size of the bumps will also increase without limit. (Figures courtesy of Rich Schwartz.)*

The general scheme of Schwartz's proof is now relatively simple to explain. Given an irrational aspect ratio $r$, you approximate it by rational numbers with odd numerators and denominators, in such a way that each approximation is "unusually close" to the next. (This is a relatively simple modification of the standard continued fraction.) Then the Arithmetic Graph for $r$ should be a limit of the Arithmetic Graphs for the approximations. Because of the Copy Theorem it is very easy to find the limit, as large chunks of the graph are simply frozen in place after a finite number of approximations. Finally, because the bumps keep getting larger and larger (like the fleas with larger fleas to hop on), the Arithmetic Graph for $r$ must move arbitrarily far from the baseline. And this, as before, means that the outer billiard for the kite with aspect ratio $r$ is unbounded.

Inspired by Schwartz's success, Dolgopyat and Fayad decided around 2008 to look at the case where the hole in the billiard table is a semicircle. They were not the first people to consider this shape: Tabachnikov had performed computer experiments with it in the early 1990s and discovered strong evidence, but no proof, of orbits that escape to infinity.

Dolgopyat and Fayad began, like Schwartz, by focusing their attention on return maps to one strip in the plane—a strip twice as wide as the semicircle, containing the flat side of the semicircle in its base. Within this strip, the return map looks like a slightly distorted version of a periodic, piecewise linear map. In Figure 7, consider the first red dot to the right of the semicircle, which lies in the infinite strip just described.

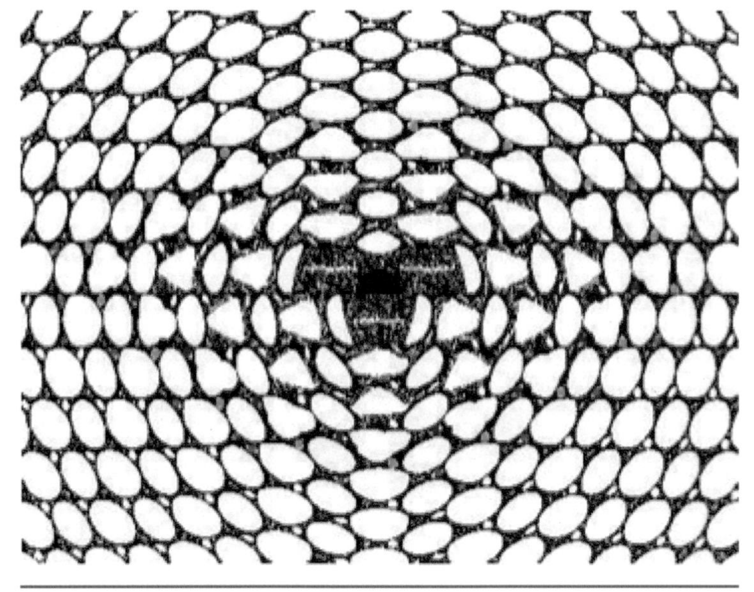

**Figure 7.** *The dynamics of an outer billiard orbit for a semicircular table (black region in center). Orbits that start in the small red or green regions head monotonically out to infinity, as shown by Dmitry Dolgopyat and Bassam Fayad. (Figure courtesy of Serge Tabachnikov.)*

**A key subtlety here is the fact that the billiard map is area-preserving. . . Thus, at every step the dot is squeezed an equal amount in one direction as it is stretched in the other direction. So neither the squeezing nor the stretching can ever predominate.**

At the center of the dot lies a point whose trajectory under the billiard map leads roughly to the center of the next dot to the right, and then roughly to the center of the dot after that, and so on. Thus this point is clearly a candidate for an orbit that escapes to infinity. Fayad and Dolgopyat wanted to prove more: if the billiard ball is placed at *any* point within the red dot (not just at the center point) it will also march steadily out to infinity.

There are two complications that could spoil the picture. First, the linear approximation to the return map does not map each dot directly over to the next dot. It stretches the dot a little bit in one direction, while compressing it in another, as if it were kneading a lump of dough. If the dot were stretched a little bit more each time, it could eventually become long and skinny enough to cross one of the discontinuities. Dolgopyat and Fayad wanted to avoid that, because it would make the rest of the trajectory much more difficult to figure out.

The saving grace is that the return map also *rotates* the dot while stretching it. So on the next iteration, the direction that was stretched previously may be compressed, and vice versa. Over the long run, the net effect is the same as kneading a lump of dough many times in different directions. It undergoes a roughly equal amount of compressing and stretching in each direction, and thus it will retain a nearly circular shape. A key subtlety here is the fact that the billiard map is area-preserving, as we mentioned in passing earlier. Thus, at every step the dot is squeezed an equal amount in one direction as it is stretched in the other direction. So neither the squeezing nor the stretching can ever predominate.

The second complication is that the return map has nonlinear distortions, attributable to the curved side of the semicircle. These distortions can nudge the center of the dot away from the place where we expect it to be. Over enough iterations of the return map, these little nudges could accumulate enough to force the dot to cross one of the discontinuities.

But again, there is a saving grace. The nonlinear distortions shrink in proportion to the distance of the billiard ball from the semicircle—because, as you get farther from the semicircle, it becomes harder to "see" its nonlinearity. (Remember, from far way it just looks like a speck.) The only way that the perturbations could get out of control would be if they all pushed the center of the dot in the same direction. (That is because the harmonic sum, $\sum 1/n$, diverges, so that a sum of small $1/n$-sized pushes can add up to a very large displacement... but only if the pushes are all in the same direction.) Dolgopyat and Fayad showed that this kind of cooperation cannot occur, for roughly the same reason as before—the perturbations actually tend to distribute themselves equally in all directions.

It is a striking fact that all the outer billiard shapes discussed so far have led to very different dynamics: periodic orbits for rational polygons, bounded orbits for uniformly convex curves, erratic orbits that are highly sensitive to the starting point for irrational kites, unbounded but stable orbits for the semicircle. It is probably too early, at this point, to talk about a general theory of outer billiards; each case seems to offer fresh surprises. "It's a simple model that exhibits deep and rich behavior," says Tabachnikov. With many interesting phenomena still to be uncovered, and with computers to guide the way, it seems likely that the "trick shots" discovered by Schwartz, Dolgopyat, and Fayad will not be the last.

**Dmitry Dolgopyat and Bassam Fayad.** *(Photo courtesy of Dmitry Dolgopyat.)*

**Figure 1.** *Until 2009, routine mammographies had been recommended on an annual basis for women over 40. Research using six mathematical models of breast cancer demonstrated rather small benefits of annual screening versus biennial screening. An independent task force of physicians recommended that women younger than 50 should no longer be counseled to undergo annual screening.*

# SimPatient

IN NOVEMBER OF 2009, a previously little-noticed panel of medical experts stirred up a political ruckus in the United States by recommending that women between the ages of 40 and 49 should no longer be counseled to undergo routine mammograms. The new guidance by the U.S. Preventive Services Task Force (USPSTF) revised the advice that the same group had issued in 2002.

Immediately politicians from both parties pounced on the news. "We can't allow the insurance industry to continue to drive health-care decisions," said Democratic congresswoman Debbie Wasserman Schultz. On the Republican side of the aisle, Representative Marsha Blackburn said, "This is how rationing begins. This is the little toe in the edge of the water. This is when you start getting a bureaucrat between you and your physician." Both sides conveniently ignored the fact that there were no insurance representatives, and no "government bureaucrats," on the task force.

The Department of Health and Human Services, which funds the task force, went into retreat mode. "The Task Force has presented some new evidence for consideration but our policies remain unchanged," read a statement issued by HHS Secretary Kathleen Sebelius. "Indeed, I would be very surprised if any private insurance company changed its mammography coverage decisions as a result of this action."

In the overheated political climate, very few people noticed one of the most remarkable features of the Task Force recommendation. The guidelines were based on mathematical models, supplementing randomized controlled trials (the previous "gold standard" of clinical evidence). For the first time, simulated patients were being used to guide the choices offered to real patients.

Why simulations, and why now? In part, because mammography has become so widely accepted. Randomized controlled studies are hard to justify on ethical grounds, once a treatment has become established. It has also become difficult, in the U.S., to ensure that the control group actually complies with the protocol and avoids screening. As a result, only one new controlled study has appeared since the USPSTF addressed the issue in 2002, and it was conducted in the United Kingdom. Most of the earlier controlled trials (one of them nearly 50 years old) do not reflect current technology and methods, and so their relevance to measuring the effectiveness of today's mammography screening is debatable.

Also, a mathematical model can "interpolate" between previously performed clinical studies. "You know that it reproduces in a reasonable way the results of the randomized trials, and you can simply change your assumption about how the women are screened," says Donald Berry of M.D. Anderson Cancer Center, who developed one of the six models in the study. "What is the effect of biennial screening versus annual? Models are the only reasonable way to address that question."

**In November of 2009, a previously little-noticed panel of medical experts stirred up a political ruckus in the United States by recommending that women between the ages of 40 and 49 should no longer be counseled to undergo routine mammograms.**

**Sylvia Plevritis.** *(Photo courtesy of Sylvia Plevritis.)*

In fact, the models in the USPSTF study compared 20 different screening strategies, many more than any controlled trial could. They examined biennial and annual screening, conducted over ten different age ranges, from ages 40-84 (the longest duration) to ages 55-69 (the shortest). For each strategy, they estimated the number of lives saved and the number of life-years gained. The models also estimated the harms, in terms of extra mammograms performed, unnecessary biopsies, and "overdiagnosis"—in other words, treatment of cancers that would never have caused clinical symptoms. Most of the models concluded that initiating routine mammograms at age 40 *would* save some lives, but surprisingly few and at a very high cost.

The models did not replace human judgement. It was still up to the task force to weigh the costs and benefits and choose the most efficient strategy. Out of the 20 strategies, the task force judged biennial screening from ages 50-74 to be most efficient. Reasonable people could, and did, disagree: for example, the American Cancer Society continues to recommend annual screening beginning at age 40. But the models did provide hard evidence to base the decision on.

Sylvia Plevritis of Stanford University, the lead developer of one of the six models, says that she was disappointed by the general dismissal of the study in the public arena. "No one acknowledged that there was a scientific process in place, and that the task force recommendations were based on the best available information."

How does a cancer model work? Each of the six models in the study (which were developed at Georgetown University, the Dana-Farber Cancer Institute, the University of Wisconsin, M.D. Anderson Cancer Center, Stanford University, and Erasmus University in the Netherlands) adopted different approaches. However, they all agreed to use shared input data to calibrate their models. These inputs included incidence rates for breast cancer, empirical data on the types of screening and treatments in use, and data on other causes of death.

The Stanford model, for instance, contains six components (see Figure 2), among which the natural history component was the most distinctive. In a clinical setting, breast cancer is usually evaluated in stages. Most of the models represented the progress of the disease using these stages. They assumed a fixed probability for the cancer to progress from one stage to the next in each time interval, depending on the patient's age and treatment style and other variables. This type of model, with discrete times and disease states, is called a Markov chain.

The trouble with a Markov chain is that not all stage II cancers are the same. As the tumor continues to grow, its chance of spreading to nearby lymph nodes and metastasizing to distant locations continues to increase. Therefore, Plevritis chose to represent the tumor's growth as a continuous function of time. Her group assumed that its volume is a simple exponential function, $V(t) = c_0 e^{t/R}$, where the parameter ($R \ln 2$) represents the time required for the tumor to double in size. Plevritis estimated the median doubling time for the whole population by calibrating the model's output against clinical and population data (it turned out that 0.75 years was the best fit). However, not every patient has the same doubling time; some will

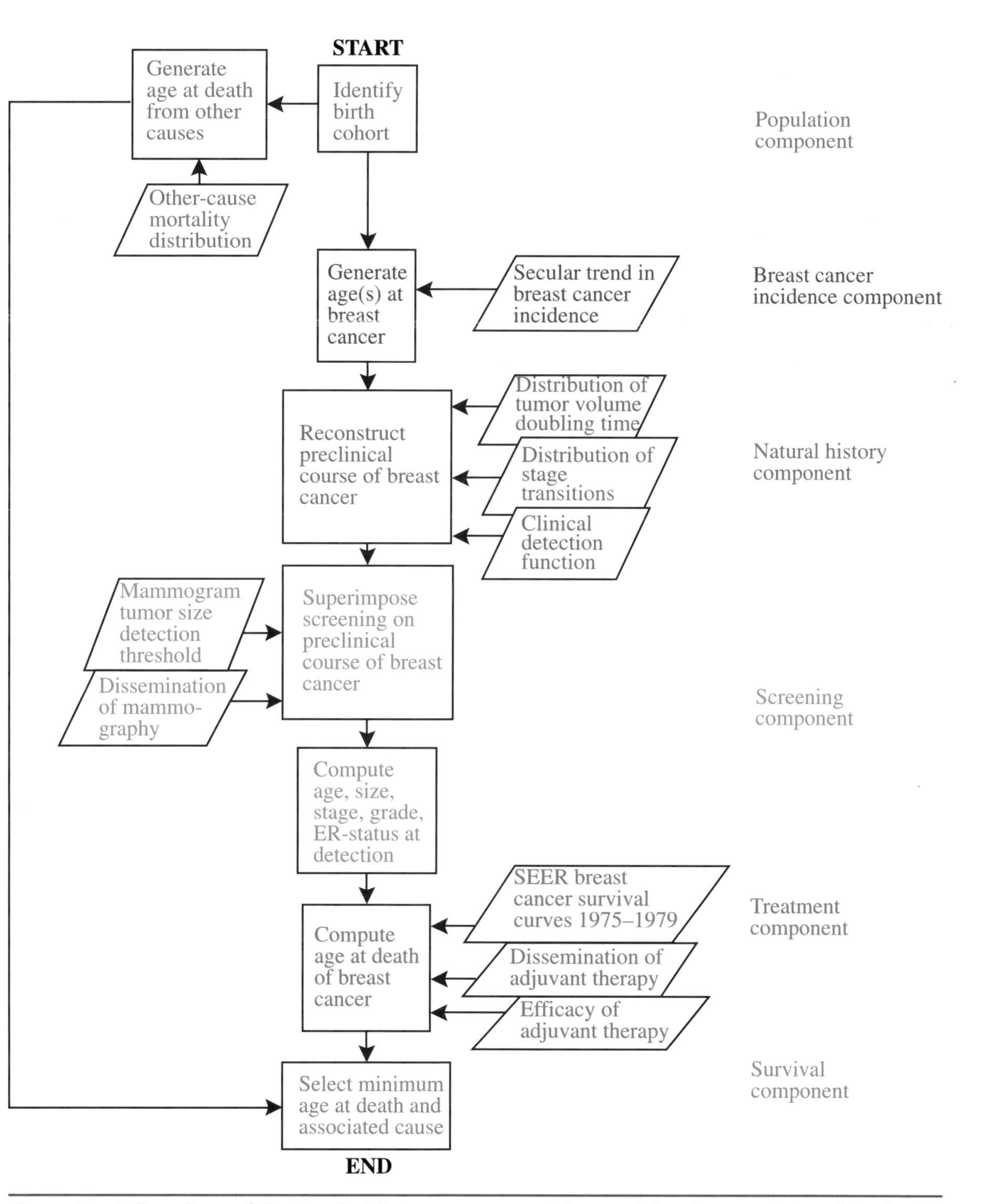


**Figure 2.** *A flow chart of the Stanford breast cancer model, one of the six commissioned by the U.S Preventive Services Task Force. The model is a Monte Carlo simulation of individual patients, rather than populations. Parallelograms represent inputs. The natural history component is a deterministic model of the physical growth of the tumor; other components are based on empirical probabilities inferred from population data. (Figure courtesy of Sylvia Plevritis and the Journal of the National Cancer Institute Monographs, 2006, pages 86-95; A Stochastic Simulation Model of U.S. Breast Cancer Mortality Trends from 1975 to 2000 (Figure 1).)*

have rapidly-growing tumors and some less rapid. Thus, for each simulated patient an individual value of $R$ was assigned at random, according to a statistical distribution known as the gamma distribution. The width and shape of this distribution are also model parameters that have to be estimated by comparing the results of the model with real data. (See Figure 3)

Although exponential growth is an unrealistic assumption at the higher end of the tumor size scale, Plevritis' group found that this did not matter very much because the cancer metastasizes long before the tumor gets unrealistically large. In the model as well as in reality, it is the dissemination of cancer throughout the body, not the original tumor, that kills the patient.

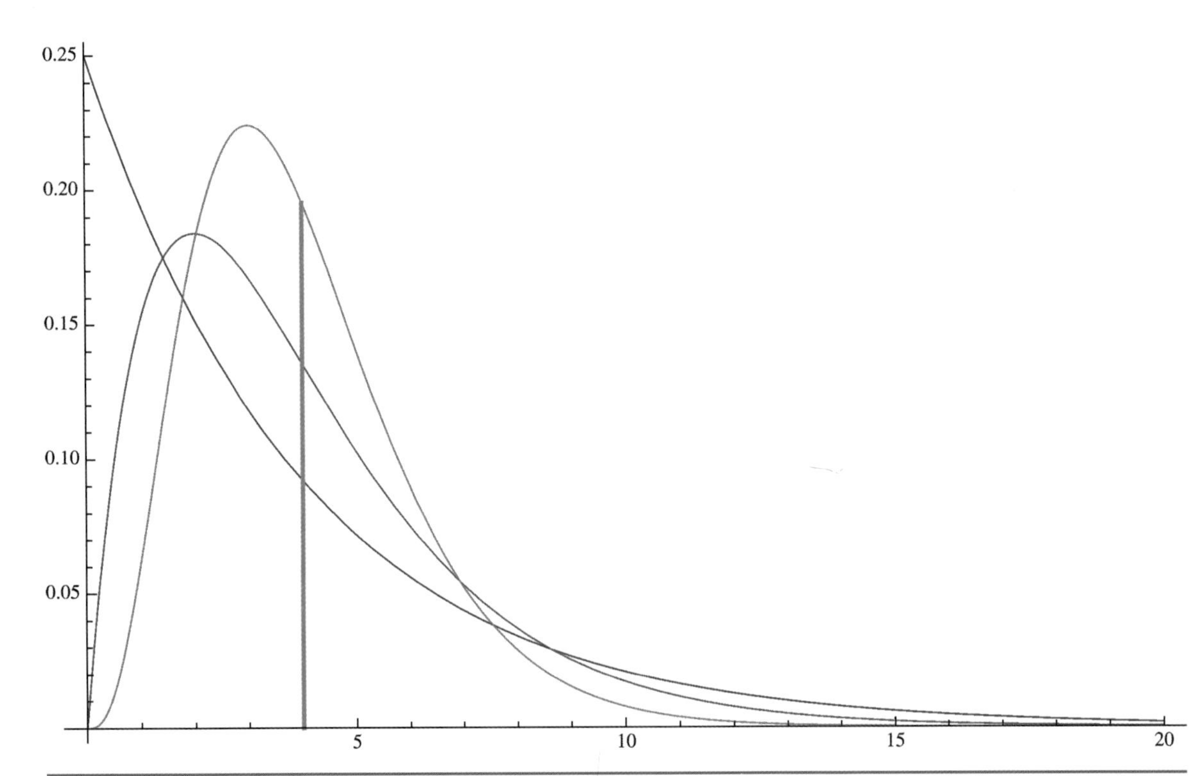


**Figure 3.** *A key parameter in the Stanford model is the doubling time of a patient's tumor. In the model, the doubling time is set for each patient by a random draw from a gamma distribution, shown here. The mean of the distribution is shown in red (about 0.75 years for the Stanford model). Gamma distributions with the same mean can come in a variety of shapes, as shown here, so the shape of the distribution also has to be inferred from data. (Figure courtesy of Chris Boucher.)*

In Plevritis' model, the probability that the cancer spreads (first from local to regional cancer, then from regional to metastatic) is described by two "hazard functions" that increase linearly as the volume increases: $P_{regional}(t) = \eta_0 + \eta_1 V(t)$ and $P_{metastatic}(t) = \omega_0 + \omega_1 V(t)$. The model also assumes that the probability of clinical detection (due to the appearance of symptoms, not due to routine screening) is directly proportional to the tumor's volume: $P_{detection}(t) = \gamma V(t)$. The parameters $\eta_0, \eta_1, \omega_0, \omega_1, \gamma$ are not directly observable, but they also can be inferred by calibrating the model against clinical data.

The next step after the natural history component is a screening component. First, each simulated patient has a screening schedule randomly assigned, according to the current clinical practice for her age and birth year. Because the detection of a tumor depends not on its volume but its cross-sectional area, the probability of detection by a mammogram has a slightly different form: $P_{screen}(t) = \lambda V(t)^{2/3}$, where $\lambda$ is another parameter to be estimated from the shared clinical data.

Once the tumor is detected in the simulated patient, either by routine screening or by the emergence of symptoms between mammograms, she passes to a "treatment component" and a "survival component" of the model. Ideally, Plevritis says, she would have preferred to base these components also on the physical growth of the tumor and its spread through the body. However, she says, "We were having trouble getting it to fit the data." In addition, they did not want to slow down the momentum of the other consortium partners. Thus, for the purposes of the USPSTF study, the Stanford team settled for a sort of hybrid, combining the physical model (described above) with empirical survival functions based on cancer stage. They have since developed an alternative version, called the "cure/no cure model," which fits the survival curves better.

After any mathematical model has been developed, it must be validated. No matter how logical it may appear on the surface, the developers still have to answer the question, "Why should anyone believe this?" This is not an idle question because no mathematical model is ever a perfect representation of reality. In the famous words of English statistician George Box, "all models are wrong, but some models are useful." For example, the assumptions of exponential growth, not to mention a spherical tumor, are clearly wrong in detail—but they may be useful in practice, and the inaccuracies may be unimportant.

**After any mathematical model has been developed, it must be validated. No matter how logical it may appear on the surface, the developers still have to answer the question, "Why should anyone believe this?"**

Validation really includes three steps: *verification* that the model works as expected for the input data that went into the model; *validation* that it accurately predicts experimental results that were not part of the assumed data, and *sensitivity analysis*, which identifies the parameters of the model that have the greatest effect. Because the "true" values of parameters like $\eta_0, \eta_1$, and so forth can never be known precisely, it is important for the modeler to indicate how much the uncertainty in these parameters affects the output of the model.

After all the groups had developed and tested their models, and performed their evaluations of the 20 different screening strategies, the final step in the process was to pool the results. This was done in a 19-author paper that appeared in the *Annals of Internal Medicine* in November 2009 and led directly to the USPSTF's decision not to recommend routine screening for women aged 40 to 49.

The *Annals* paper is interesting both for what it does and what it does not do. For each of the models, it presents an "efficiency plot," (see Figure 4, next page) showing the mortality reduction as a function of the number of mammographies performed. Efficient strategies are the ones that lie on the frontier of these plots, shown as a boldface line. Strategies that lie below

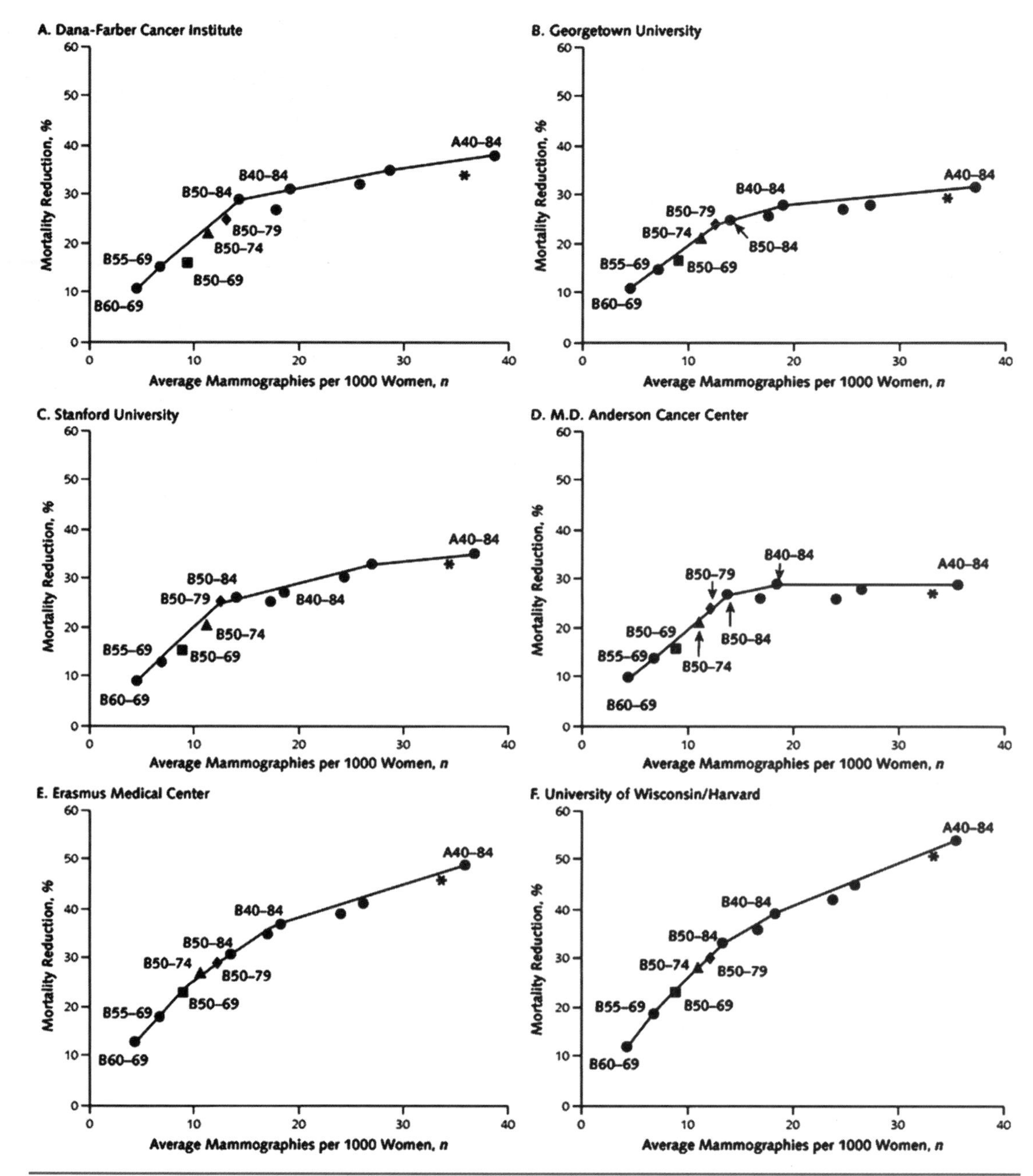


**Figure 4.** *An analysis of different screening strategies, according to the six different models. Dots indicate different strategies (e.g., B50-79 represents biennial screening for women from ages 50-79). "Frontier points" on the heavy lines represent strategies that are not dominated by any other strategy. In all six models, biennial screening strategies predominate, with the exception of A40-84 (annual screening for women from ages 40-84). The panel concluded that the mortality benefit of the A40-84 strategy did not justify the harms in terms of total mammographies (shown here), unnecessary followup procedures performed (not shown) and overdiagnosis (not shown). (Figure courtesy of the Annals of Internal Medicine, Figure page 742: Jeanne S. Mandelblatt, et al, Effects of Mammography Screening Under Different Screening Schedules: Model Estimates of Potential Benefits and Harms, Annals of Internal Medicine, November 17, 2009, volume 151).)*

the frontier are said to be "dominated" because an alternative strategy will produce a greater mortality reduction at no greater cost. Examining these graphs, in every model you can see an elbow around the points labeled "B50-79" or "B50-84." These points correspond to biennial screening either from ages 50-79 or 50-84. Extending the screening over a longer period of time, or going to annual screening (the strategies labeled "A"), yields a relatively modest improvement in mortality at a much higher cost in the number of mammographies. In one model (the M.D. Anderson model) the "A" strategies actually produce no improvement at all in mortality. Because of the predominance of "B" strategies on the frontier, the authors concluded that biennial screening was more efficient.

At the same time, the *Annals* article makes some surprising and revealing omissions. First, in every model, "A40-84" (that is, annual screening from ages 40 to 84) is a dominant strategy. Even though its cost is the highest in terms of procedures performed and false positives, it also (unsurprisingly) detects the most cancers. The report gives no clear reason for discarding it, at least for women or health plans or employers who are willing to pay the cost.

Second, although many people interpreted the study as a cost-benefit analysis, there is actually no mention of costs anywhere in the paper. To some extent, that omission is deliberate. According to David Eddy, a veteran of the breast-cancer-screening wars, "Any organization setting a national policy, such as an insurance company or a government body, that explicitly takes cost into account and recommends against screening because of cost would be lynched. Rationing would be yelled around [as indeed it was—DM] and they would be skewered."

Nevertheless, such information is available from models, and has been used for less politically charged decisions. Plevritis, for example, has used the Stanford model to compute the cost in dollars for each additional "quality adjusted year of life" (QALY), to compare different screening alternatives for women with a genetic mutation that puts them at high risk for breast cancer. Similarly, Eddy has performed cost estimates per QALY for Kaiser Permanente, a health insurance company (see Box, **"Modeling the Whole Patient**, p. 70). The lack of explicit cost estimates makes it hard to pin down the reasons for the USPSTF recommendation. "It wasn't clear to me what line they were drawing in the sand," says Plevritis, who was listed as a co-author. "There wasn't a metric." Plevritis says that the "line in the sand" could have been drawn more clearly if the participants had also simulated cancer screening for 30–39 year old women, for whom conventional wisdom says the benefits of routine screening are insufficient. "That would give us a better sense of how the benefits and harms might change," Plevritis says.

A third omission from the *Annals* report is the lack of any estimate of overdiagnosis and overtreatment. "My guess is that the models came out with totally different estimates," says David Spiegelhalter, a professor in the public perception of risk at the University of Cambridge. "I think it was a real shame that they did not put the estimates out front." The report said only this: "Of the 6 models, 5 estimated rates (Continued on p. 73)

**David Eddy.** *(Photo courtesy of Archimedes Inc.)*

### Modeling the Whole Patient

In 2002 Kaiser Permanente, a major managed care organization, wanted to evaluate the effectiveness of a new treatment regimen for patients with diabetes or heart disease. The hypothesis was that a combination of three generic drugs—aspirin, a cholesterol-lowering drug (lovastatin), and a blood pressure-lowering drug (lisinopril)—would lower the number of heart attacks and strokes in this high-risk group. The company's main objective was to prevent complications such as heart attacks and strokes. A not-so-secret side effect would be to save money; as the old adage goes, an ounce of prevention is worth a pound of cure.

Traditionally, the way to test the effectiveness of a drug is a randomized controlled study, where one group of volunteers gets the drug and the other does not. (No one knows who is getting the drug and who is getting a placebo). But how can you do this for three drugs at once, especially when one of them is found in every home medicine cabinet?

"You would not be able to do a controlled trial—the number of people, amount of money, established practice patterns, and time make it infeasible," says David Eddy, the founder of a company called Archimedes, Inc. "It involves three different drugs, three different risk factors, and several different types of patient outcomes." Instead, Kaiser Permanente (which owns Archimedes, Inc.) asked Eddy to do a simulated clinical trial.

"We calculated that the combination should reduce the number of heart attacks and strokes by 71 percent," Eddy says. "It should save about $350 per person with diabetes per year in the early years, and more in later years because you would be preventing late-stage renal disease. If you've got 500,000 or so patients with diabetes, that is a lot of money."

Kaiser Permanente was convinced: It began recommending the A-L-L combination to all of its participating doctors. A followup study in 2009—this time an observational study of real people—showed that Eddy's Archimedes Model had been essentially right. Among patients who had stuck to the lovastatin and lisinopril parts of the regimen (the use of aspirin could not be tracked), there had been a 60 percent decrease in strokes and heart attacks.

The A-L-L study is an example of the brave new future envisioned by Eddy, in which mathematical models will allow physicians to answer many more "what if" questions. Simulations can be cheaper than clinical trials, more flexible, and—dare one say it?—more realistic. Models can take into account all sorts of real-world complications, such as patient non-compliance and the infinite variety of patient histories, which clinical trials either ignore or assume to be irrelevant because of the randomization procedure.

The Archimedes Model is surely the most ambitious mathematical model of the human body ever conceived. Instead of simulating one organ or one disease at a time, it uses differential equations (hundreds of them) and object-oriented programming to mimic major parts of the human body—with a long-term goal of mimicking all of it. Eddy, together with particle physicist Len Schlessinger, has been developing the model since the early 1990s. Eddy has a long history of developing mathematical models for medicine, starting in the 1970s when he worked with Markov chain models. His models led to several landmark recommendations by the American Cancer Society in 1980, such as giving Pap smears to women every three years rather than every year, colon cancer screening every five years, and not screening smokers for lung cancer. But over the years, Eddy became disenchanted with the limitations of Markov chains—the fixed time steps, the lumping of very dissimilar patients into broad categories, the simplistic transition probabilities. He wanted to develop a model that more closely represented what was actually going on in a patient's body.

"The Archimedes Model tries to capture what you learn in medical school about all the interactions in the body," says Lynn Etheredge, an independent health policy specialist (formerly with the Congressional Budget Office). "It's an avatar model, which means that it creates tens of thousands of individuals, each with their own individual physiology, and then ages them over decades, modeling their interactions with the health system and treatments."

"The model lets you go down into the biology or up into the health system," Eddy says. This gives you flexibility to use it in different ways. A government regulator or an insurance company can estimate costs and benefits of different policies. Physicians might, in the future, consult the model for individualized medicine, to determine which drugs or combinations might work best for their patients. A harbinger of this future, called Diabetes PHD, can already be found at the website of the American Diabetes Association. Based on the Archimedes Model, Diabetes PHD allows you to input your basic medical information (height, weight, cholesterol levels, etc.) and find out your personal risk for various outcomes (diabetes, heart attack, etc.) You can easily check how, for example, a change in your smoking habits might change your risk.

Given the difficulty of getting a model of even one disease right, it's natural to be skeptical of a program that aims, eventually, to model everything. In fact, the Archimedes Model is nowhere near its destination yet. It includes cardiovascular disease, asthma, three cancers, and diabetes (see Figure 5, next page), but does not, for example, include Alzheimer's disease.

Eddy's way of answering the skeptics is to use the model to predict the result of clinical trials. "For a number of trials, he has given sealed predictions," says Etheredge. Indeed, the match between his sealed predictions and the actual results is so good that it is almost spooky (see Figure 6, page 73). The model has also retrospectively "predicted" the results of 40 other clinical studies, with a correlation of about 0.98 between the model results and the clinical trial results in the first round of validations.

Eddy stresses that the Archimedes Model will never replace lab experiments and clinical trials. For one thing, the equations include only what is known about human biology, not what is unknown. The model missed on one study of a statin that unexpectedly proved to be more effective against strokes than other statins. "But so did everyone else," says Etheredge. "It showed that they had to go back and understand the basic science."

However, the computer model *can* synthesize existing knowledge, in many ways better than a human physician. It can sometimes identify mistaken assumptions that physicians make—just because they are human. Most importantly, it can give the practitioner quantitative answers instead of gut feelings.

"Medicine needs mathematics in the same way that every other field of endeavor does," Eddy adds. "You don't build a bridge without a model, and you don't even calculate a mortgage without one. But medicine has been the holdout. We need to help this profession, which thinks qualitatively, to think more quantitatively. As the models get better validated and predict things that actually happen, modeling will play a more important role."

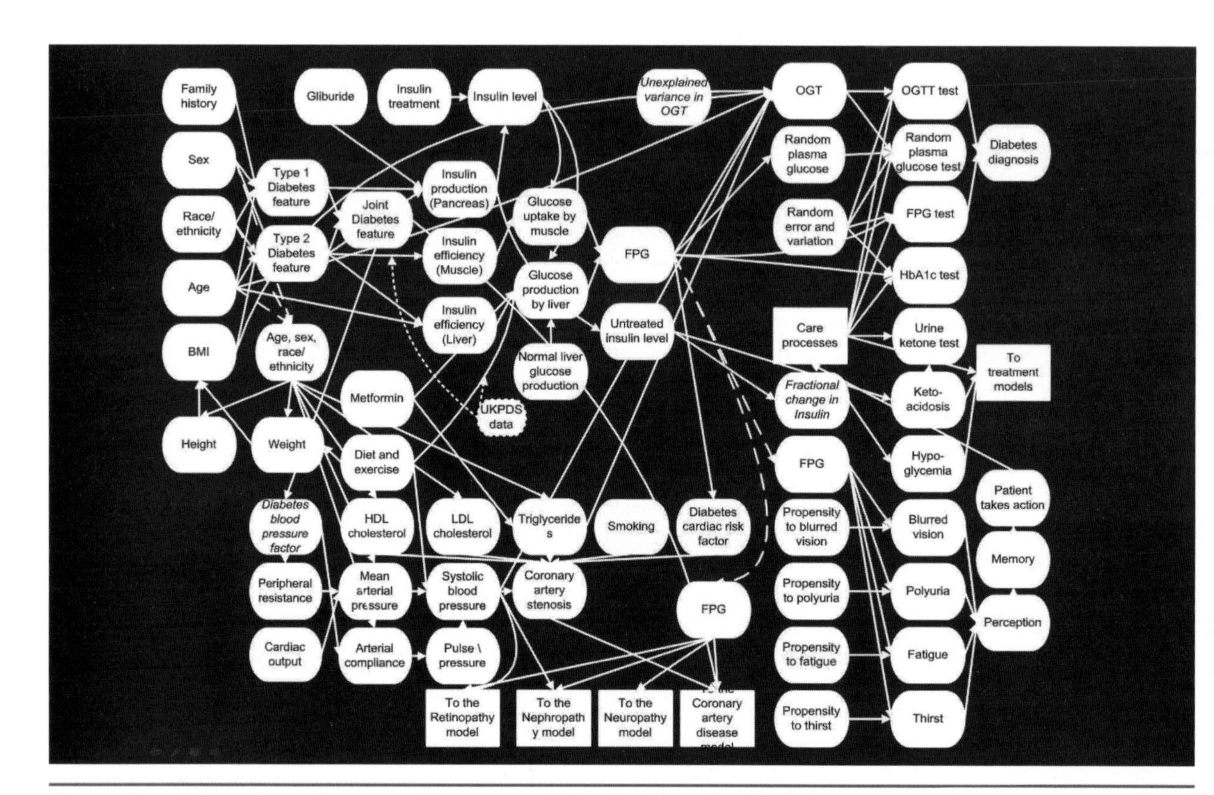


**Figure 5.** *A flowchart diagram of one part of the Archimedes Model, concentrating on systems related to diabetes. (Copyright 2003 American Diabetes Association, From Diabetes Care, Vol. 26, 2003; 3093–3101. Reprinted with permission from the American Diabetes Association.)*

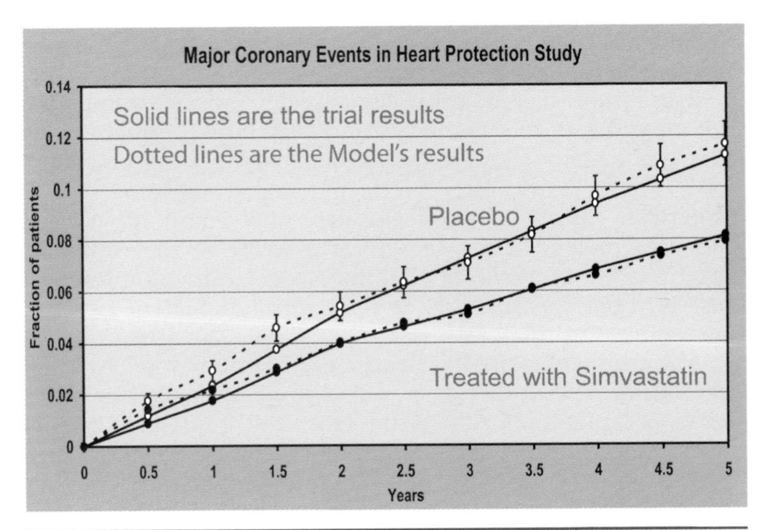


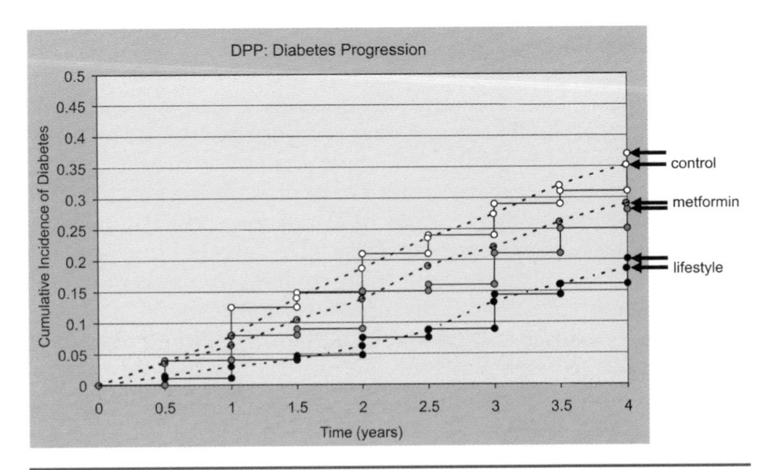


**Figure 6.** *Top: Comparison of Archimedes Model predictions to a clinical trial of a drug to prevent heart attacks in high-risk patients. Bottom: Comparison of Archimedes Model predictions to a clinical trial of two diabetes prevention programs. In both cases, the Archimedes predictions were generated and sealed before the clinical trial was completed. (Figures courtesy of Archimedes Inc.)*

(Continued from p. 69) of overdiagnosis. They showed an increase in the risk for overdiagnosis as age increases (data not shown)." Other harms of cancer screening, such as the rate of false positives, were explicitly calculated. Yet surveys have shown that most women are glad to put up with the anxiety resulting from a false positive, in order to reduce their cancer risk.

Mathematical models should be able to estimate the rate of overdiagnosis, but the problem is that the modelers do not agree on how to define it. Roughly speaking, it is the diagnosis and treatment of cancers that are either very slow-growing, or not of an invasive form, or which occur in patients who would have died of other causes first. For example, there is a condition called ductal carcinoma in situ (DCIS) that some doctors consider a precursor of breast cancer, but which others

**In sum, the breast cancer controversy suggests that mathematical models have earned a place at the table in discussions of health policy, but they are not yet trouble-free.**

consider to be only a risk factor (in the same way that smoking is a risk factor for lung cancer).

The importance of overdiagnosis depends very much on whom you talk to. "The trials and very large population series don't really show that overdiagnosis is a big problem," says Robert Smith, the Director of Cancer Screening for the American Cancer Society. On the other hand, physicians Steven Woloshin and Lisa Schwartz wrote in the *New England Journal of Medicine* after the cancer screening tempest broke, "false positives are not the most important harm of screening—overdiagnosis is."

The problem is perhaps most easily understood by comparison with prostate cancer, a slow-growing cancer for which most organizations that issue screening guidelines do not recommend routine screening. The reason is that most men with prostate cancer will die of something else before they die of prostate cancer. Moreover, the harms of treatment are substantial. "For every life saved, many men are rendered incontinent and impotent," says Spiegelhalter. "That's a pretty heavy risk balance."

For breast cancer, the balance is not so extreme, and the proper choice is not so clear. That is why a properly designed mathematical model could have been informative, if the information were made public. It is clear that *some* breast cancers are very slow-moving; the tumors are present for years before any symptoms appear. Some may not ever kill the patient. Unfortunately, screening preferentially detects slow-growing cancers, while missing some more aggressive cancers, which can proceed from the microscopic level to the metastatic level between scheduled screenings. In the *New England Journal of Medicine* (*NEJM*), a month *before* the controversy broke, surgeon Laura Esserman of the University of California at San Francisco even recommended a change in terminology, referring to slow-growing cancers as "IDLE tumors" rather than "cancers." (IDLE is an acronym for *indolent lesions of epithelial origin.*)

Esserman estimated that as many as one out of every three breast cancer patients will undergo treatment for a disease that would not have killed her. The harms are substantial and undeniable: potentially disfiguring surgery, life-disrupting radiation and chemotherapy. The trouble is that you will never know *which* one-third of the cancers (if that estimate is correct) could have been left alone. No woman will ever tell you, "I was overdiagnosed"—she will always say, "My life was saved." Only a mathematical model, with simulated patients, can ever tell which women are right and which are wrong.

Overdiagnosis is the primary reason why the USPSTF recommended *stopping* routine mammograms at age 74—a part of the report that received much less scrutiny in the media than the recommendation to *start* at age 50. It's not because the risk of breast cancer declines after age 74—in fact, the opposite is true. The older you get, the greater your risk of developing breast cancer becomes. But your risk of dying from some other cause also increases dramatically after age 74, so the benefit of detecting and treating the cancer is either small or nonexistent. On the other hand, a 75-year-old woman in excellent health might be well-advised to continue breast cancer screening.

This is a final point for a confused member of the public to bear in mind. The *Annals* study, as Smith says, "paints with a broad brush." It is a study written for policy-makers, not for individual patients. The task force made this clear in its report: "The decision to start regular, biennial screening mammography before the age of 50 years should be an individual one." However, this nuance was for the most part lost in the political shouting that ensued.

**However, the acceptance of mathematical models for medical decision-making—at least behind the scenes—continues to grow.**

In sum, the breast cancer controversy suggests that mathematical models have earned a place at the table in discussions of health policy, but they are not yet trouble-free. They did not, for example, provide sufficient consensus on the important question of overdiagnosis. They are also opaque to the outsider. (A significant part of the population does not understand randomized controlled trials, for instance, and they are much easier to explain than a mathematical simulation.)

"What concerns me is that models are seductive," says Smith. "The average person really can't ascertain what's taking place in the model. It's not clear how the model inputs are selected. Is it because they fit the personal preferences of the modeler?" In fact, Smith believes that the biases of the researchers did affect some of the breast cancer models: "To imagine that opinions and values don't enter into this process is fanciful."

On the other hand, Smith does think that models are useful, particularly for sensitivity analysis. For example, if we can improve the quality of our mammograms by $x$ percent, how much can we decrease the screening frequency? He collaborated with David Eddy, the creator of the Archimedes Model, on a study of exactly this question.

Plevritis thinks that the models could have been used to make a more nuanced recommendation. "There are women at different levels of risk," she says. "They could have given advice that was more patient specific." Though the task force left the decision on starting mammography in the 40s up to the patient and her physician, it did not provide any concrete guidelines on how to make this decision.

Of course, the controversy over the panel's finding had nothing to do with the use of mathematical models. The debate over breast cancer screening has a very long history, and it has a track record of eliciting strong emotions. In the battle for objectively based standards of medicine, emotions may have won this time. However, the acceptance of mathematical models for medical decision-making—at least behind the scenes—continues to grow.

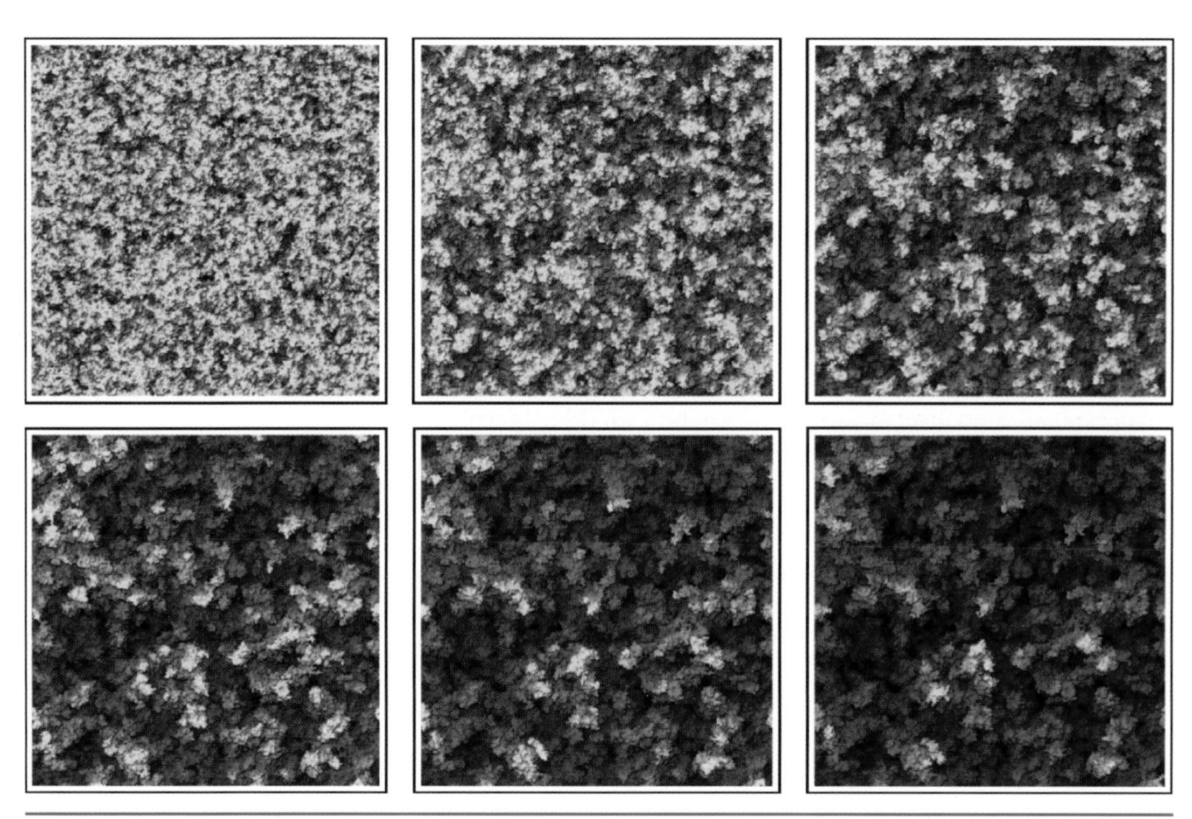

***Fading Memory.** In Eyal Lubetzky and Allan Sly's proof of the cutoff phenomenon in the Ising model, a key point is that very few of the pixels in the initial configuration actually affect the final configuration at the mixing time. Here, light lattice points at time $T$ represent the pixels in the initial configuration that are still relevant at time $T$. (The darkness of the other pixels represents how long ago that pixel's value in the initial configuration became irrelevant.) (Figures courtesy of E. Lubetzky and A. Sly.)*

# Instant Randomness

**Eyal Lubetzky.** *(Photo courtesy of E. Lubetzky and A. Sly.)*

HOW MANY SWIPES OF YOUR SPOON will it take to stir your milk into your coffee? How many shuffles of a deck of cards does it take to randomize them? In many systems like these, the onset of randomness turns out to be quite sudden. Mathematically speaking, the system goes from unmixed to mixed in the blink of an eye.

The name for this abrupt mixing behavior is the "cutoff phenomenon," and the time $T$ when mixing occurs is called the mixing time. To be more precise, when you stir your coffee or shuffle your cards a little bit less than the prescribed amount of time, your system is still a long way from being mixed. If you go a little past the mixing time, your system is essentially completely mixed, and further stirring or shuffling will not make very much difference.

The cutoff phenomenon was first observed in the 1980s, in some models chosen more for their mathematical tractability and symmetry than their relevance to physics. But recently, a team of mathematicians at Microsoft Research showed that cutoff occurs in several of the most important models of statistical physics, such as the Ising model (which simulates ferromagnetism), the Potts model, and the hard-core gas model. The result, announced by Eyal Lubetzky and Allan Sly, lends plausibility to a broad conjecture about mixing made a few years ago by Yuval Peres, also of Microsoft Research.

Peres has conjectured that cutoff is a general feature of systems in which the time for a small, local deviation from randomness to become smoothed out (called the "relaxation time") is less than the time for the entire system to achieve randomness (called the "mixing time"). Intuitively, the idea behind Lubetzky and Sly's work is this: if relaxation happens faster than mixing, then small deviations from randomness in different locations (such as small patches of milk in your coffee) do not have time to communicate with each other. Thus they evolve almost independently, and they are governed by the statistical laws of independent processes. In a nearly psychic way, once one patch becomes randomized, nearly all of them do. "My feeling, and conjecture, is that cutoff represents what happens to temporal dynamics when you have relative spatial independence," says Peres.

"When I first heard about Peres' conjecture, I said, 'Give me a break.' I didn't believe it," says Persi Diaconis of Stanford University. "Now we're proving it in example after example."

Although Peres' hypothesis remains unconfirmed in general, for the Ising model—perhaps the most thoroughly studied system in statistical physics—Lubetzky and Sly's theorem is "hard and original and definitive," Diaconis says. The cutoff phenomenon occurs in every case where you can reasonably expect it, and their work determines the cutoff time $T$ precisely.

**Allan Sly.** *(Photo courtesy of E. Lubetzky and A. Sly.)*

## Simple Yet Confounding

The Ising model is named after Ernst Ising, who wrote about it in his 1924 doctoral thesis; sometimes it is called the Lenz-Ising model, because Ising's dissertation advisor Wilhelm Lenz had described it earlier in 1920. It is one of the simplest examples of a physical system that exhibits a phase transition. Even though it does not literally describe what happens in a magnet, it has served physicists well as an analogue or paradigm of the ferromagnetic phase transition.

In the Ising model, the magnet is considered to be a lattice of point atoms, each one of which can have a "spin" of $+1$ or $-1$. Neighboring atoms in the lattice like to have the same spin, and so an energy penalty accrues for each neighboring pair with opposite spins. For simplicity, the energy of a matching pair can be assumed to be $-1$ and the energy of a pair with opposing spins is $+1$. The energy of an entire configuration of spins $\sigma$, denoted by $H(\sigma)$, is the sum of the energy of all the adjoining pairs:

$$H(\sigma) = -\sum_{u\sim v} \sigma(u)\sigma(v).$$

(Here the sum is taken over all pairs of adjacent vertices in the lattice; $u \sim v$ signifies that the vertices $u$ and $v$ are adjacent; and $\sigma(u)$ is the spin at vertex $u$, either $+1$ or $-1$. Note that $\sigma(u)\sigma(v) = 1$ when the spins at $u$ and $v$ match.)

Ising and Lenz assumed that the probability of any particular configuration of spins, $\mu(\sigma)$, decreases exponentially as the energy of the configuration increases. High-energy configurations are not impossible; they are merely unlikely. In symbols, their assumption can be written as follows:

$$\mu(\sigma) \propto e^{-\beta H(\sigma)}.$$

Two points should be made about this equation. First, the proportionality constant $\beta$ plays the role of an *inverse temperature*. When $\beta$ is large (i.e., the temperature is near zero), high-energy configurations are very improbable. In such a state, the system should have a very strong preference for having all the atoms with the same spin. As $\beta$ decreases (or the temperature increases), configurations with high energy $H$ become more likely. This means that the system can tolerate a lot of non-matching spins, and there is more disorder—as we would expect at a higher temperature. Finally, when $\beta = 0$ (or the temperature is infinite), the system is completely disordered; every configuration has an equal probability of occurring.

The second point is that the exact probability density function $\mu$, called the Gibbs distribution, is beyond the power of any computer, now or in the future, to compute. The reason is that $\mu$ has to be normalized so that the sum of all the probabilities is 1. To do this, you would have to compute the weighting factor $e^{-\beta H(\sigma)}$ for every single configuration $\sigma$, and then divide by the sum. (This sum is called the "partition function.") How many configurations are there? Even for a tiny, 10-by-10 lattice, there are 100 vertices and $2^{100}$ different ways to assign a spin to each of them. To store all of this information you would need a computer the size of the universe. Now imagine trying to do the same thing for a billion-by-billion-by-billion lattice, representing a real-world crystal!

As a descriptive model for magnets, the Ising model has been very successful. (See Figure 1, next page.) It predicts a sharp phase transition, at a temperature that is known for the 2-dimensional lattice (it occurs when $\beta = \frac{\ln(1+\sqrt{2})}{2} \approx 0.44$ or at a "temperature" of $1/\beta \approx 2.27$; this of course is an abstract number rather than an actual temperature in kelvins). The critical temperature is still unknown for the 3-dimensional lattice. Above the critical temperature, disorder wins: the spins of distant lattice sites are very weakly correlated, and the total magnetization (the sum of all the spins) is likely to be close to zero. Below the critical temperature, order reigns: lattice points that are far away from each other are likely to have the same spin, and the total magnetization is likely to be close to $+n$ or $-n$ (where $n$ is the number of atoms in the lattice). In real magnets, this transition occurs at a temperature called the Curie point. For example, iron will not magnetize above 768° K. but will magnetize below that temperature.

**Yuval Peres.** *(Photo courtesy of G. Bergman.)*

Unfortunately, the Ising model is hard to analyze because the Gibbs distribution is, for all practical purposes, uncomputable. This is a common problem in statistical mechanics. For example, the designers of the first atomic bombs wanted to understand how a typical arrangement of neutrons would behave in a chain reaction. But how do you tell what is "typical" when you cannot compute the probability of any particular arrangement?

## Markov Chain Monte Carlo

In 1953, Nicholas Metropolis, a physicist at Los Alamos National Laboratories, along with four collaborators (including Edward Teller, the designer of the hydrogen bomb) discovered a remarkably general solution to the problem of taking a representative sample from an uncomputable probability distribution. His technique, called the "Metropolis algorithm," was named one of the ten most important algorithms of the $20^{th}$ century by *Computing in Science and Engineering* magazine. It was the first of a general class of methods now known as Markov Chain Monte Carlo, or MCMC for short.

Let's see how the Metropolis algorithm works for the Ising model. Remember that the goal is to generate random configurations of spins, in such a way that the probability of generating any configuration $\sigma$ is equal to the unknown probability function, $\mu(\sigma)$.

We start off with an arbitrarily chosen configuration of spins, $\sigma_0$. Note that "arbitrary" does not mean "random." In fact, $\sigma_0$ is likely to have a very high energy, and therefore to be very improbable. But we don't let that bother us. Metropolis now says to pick a lattice site at random and consider the new configuration $\sigma_1$ generated by changing the spin at that one vertex. If $\sigma_1$ has less energy than $\sigma_0$, then we accept the change. (Lower energy is always good.) If the proposed change will increase the energy, we still accept the change some of the time. To be precise, we accept it with a probability equal to $\mu(\sigma_1)/\mu(\sigma_0)$. In other words, higher energy isn't always bad!

We repeat this process over and over, typically a few thousand times in practice. As a general rule the configuration will drift down to a lower and lower energy state, but the occasional

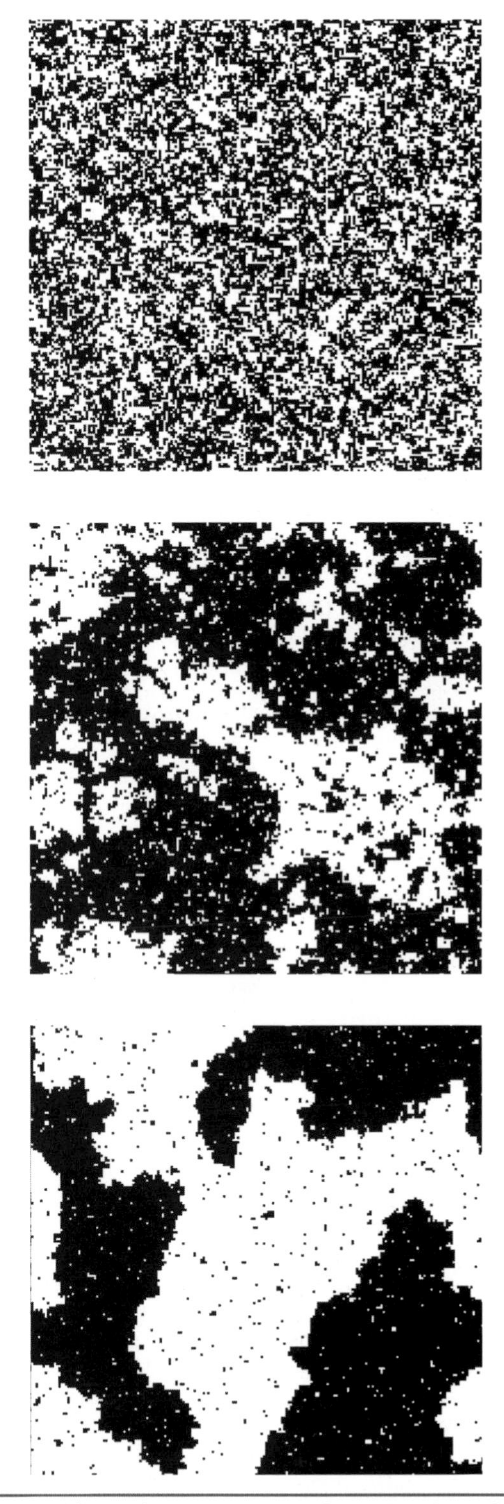

**Figure 1.** *Markov Chain Monte Carlo (MCMC) methods are a way of producing samples from a very complicated and non-uniform probability distribution, such as these "typical" configurations of the Ising model on a square lattice. (a) A typical pattern at high temperature; (b) a typical pattern at the critical temperature; (c) a typical pattern at low temperature. (Images courtesy of Matthew Richey.)*

increases will keep it from getting stuck forever in a minimum energy state. These two trends balance each other perfectly, so that the probability that $\sigma_n = \sigma$ (for a sufficiently large $n$) is roughly equal to $\mu(\sigma)$. In other words, $\sigma_n$ is essentially a random configuration.

One thing may seem puzzling about this algorithm: How can we compute $\mu(\sigma_1)/\mu(\sigma_0)$ if we don't know the Gibbs distribution $\mu$? The answer is that the uncomputability of $\mu$ is entirely due to the normalization constant, the "partition function" mentioned earlier. But if we are computing a *ratio* of probabilities, we don't need to know the normalization constant. Not only that, a look at the definitions of $H$ and $\mu$ makes it apparent that this ratio is very simple to compute; it depends only on the spins at the four (in 2 dimensions) or six (in 3 dimensions) vertices that are adjacent to the given vertex. Thus, to implement each step of the Metropolis algorithm, we need only *local* information about our configuration.

The Metropolis algorithm is nevertheless guaranteed (given enough time) to produce samples that have the right *global* probability distribution, $\mu$. This is a consequence of a general theorem about Markov chains, which was known long before Metropolis.

A Markov chain is a random system that has a finite number of states (See Figure 2 for a simple example. In practice, the number of states may be very large—$2^{100}$ in the case of a 10-by-10 Ising lattice.) At each time step the system moves from one

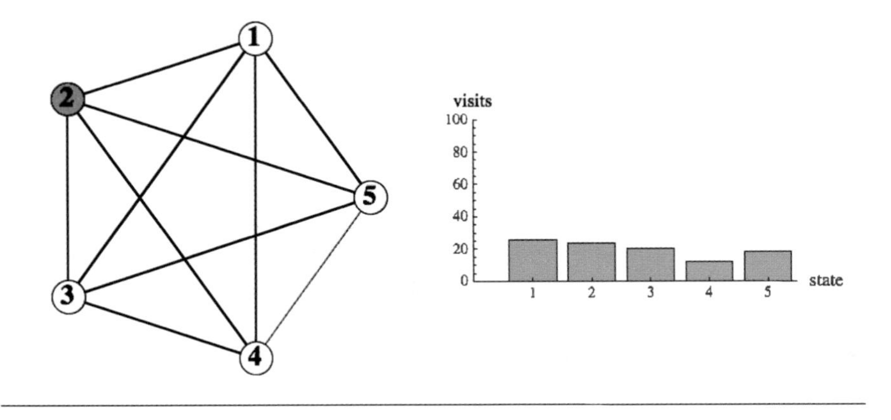


$$\begin{pmatrix} 0.165 & 0.323 & 0.293 & 0.193 & 0.026 \\ 0.330 & 0.258 & 0.048 & 0.247 & 0.118 \\ 0.093 & 0.215 & 0.428 & 0.033 & 0.230 \\ 0.364 & 0.281 & 0.066 & 0.052 & 0.238 \\ 0.341 & 0.064 & 0.347 & 0.015 & 0.233 \end{pmatrix}^{10} = \begin{pmatrix} 0.237 & 0.235 & 0.250 & 0.121 & 0.157 \\ 0.237 & 0.235 & 0.250 & 0.121 & 0.157 \\ 0.237 & 0.235 & 0.250 & 0.121 & 0.157 \\ 0.237 & 0.235 & 0.250 & 0.121 & 0.157 \\ 0.237 & 0.235 & 0.250 & 0.121 & 0.157 \end{pmatrix}$$

**Figure 2.** *A five-state Markov chain. The matrix at left represents the probability of moving from any particular position to any other position at any given step. (For example, the probability of moving from position 1 to position 2 is 0.323.) Even after as few as 10 steps, the probability of landing at any position is essentially independent of the position you started at, because all five rows of the transition matrix on the right are identical. The histogram at upper right shows the probability of being in any given location after 10 steps. This rapid convergence to an equilibrium probability distribution is called "rapid mixing." (Figure courtesy of Chris Boucher.)*

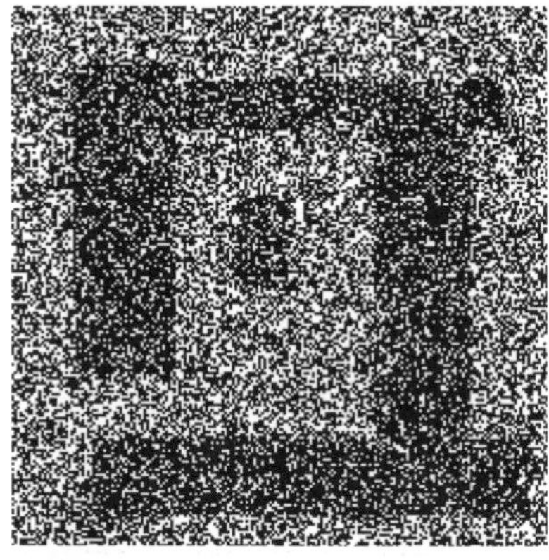

**Figure 3.** *An application of MCMC to image analysis. (a) Original image. (b) Image with white noise added. (c) Image reconstructed by MCMC. The algorithm tries to match the data in (b), but with the assumption that adjacent pixels do not like to have different colors. The model used is equivalent to the Ising model with a background "magnetic field." (Images courtesy of Matthew Richey.)*

state (say $\sigma_0$) to another state (say $\sigma_1$) with a fixed transition probability. Another way of saying this is that a Markov chain has no "memory" of where it has been before. It only knows where it is now. If the same state $\sigma_0$ comes up 1000 steps later, the system will still have exactly the same probability of moving to state $\sigma_1$.

Markov chains always (subject to some mild hypotheses) converge to a unique *stationary* distribution. That is, over the long run, there is a unique probability distribution $\mu^*$ such that every state $\sigma$ will be visited $\mu^*(\sigma)$ percent of the time. For the Ising model $\mu^*$ is uncomputable, because it would require finding the eigenvectors of a $2^{100}$-by-$2^{100}$ matrix. But it is easy to check that the (also uncomputable!) Gibbs distribution $\mu$ is stationary for Metropolis's Markov chain. Therefore, because the stationary distribution is unique, we can be certain that $\mu$ and $\mu^*$ are the same even though we cannot compute either one!

Metropolis' invention touched off a slow-motion revolution that has by now moved into nearly all areas of science; Diaconis calls it the "Markov Chain Monte Carlo revolution." For example, it has been used in image analysis (see Figure 3) and simulations of the economy.

One of Diaconis' favorite examples came about when a prison official brought an encrypted text to the Stanford statistics department. The message had been encrypted by a prisoner, using a system of symbols that does not resemble the English alphabet. (See Figure 4) Diaconis' student, Marc Coram, used the Metropolis algorithm to decipher the message.

Here is how Coram did it. In this case, he considered a "configuration" to be a proposed decryption scheme, in other words a rule that matches each symbol in the cipher text to an English letter. The plausibility of each proposed decryption scheme could be measured by multiplying the probabilities of each letter pair in the deciphered message. For instance, the letter pair "QU" is more likely to appear in an English text than the letter pair "QZ".

The Metropolis algorithm would propose one change at a time. For example, the proposal might be to change every "E" in the deciphered text to an "I" and vice versa. If the resulting text became more plausible as a result, the change would be accepted. But, as is normal with the Metropolis algorithm, even a change that produced a less-plausible deciphering would still be accepted sometimes, with a probability given by the ratio of plausibilities.

One would expect that, over time, the system would move towards more and more plausible decryptions. In fact, after a few thousand steps, it came up with the following decryption: "... to bat-rb. Con todo mi respeto. I was sitting down playing chess with danny ..." Note that the algorithm did not assume any knowledge of the English language—in fact, an ordinary cryptologist might have been thrown off by the Spanish language and prison slang that was mixed in with the English. The only assumption was that the probability distribution of letter pairs in the decrypted text would be similar to English, which (luckily) turned out to be true.

## But There's a Catch . . .

Although MCMC methods are versatile, there is one problem that bedevils many applications. The theory of Markov chains guarantees that the simulated probability distribution will converge to the actual stationary distribution. But how long does it take? A hundred steps, a thousand, a million? The period of time before the Markov chain converges to a reasonable approximation of the target distribution is known in practice as the "burn-in period."

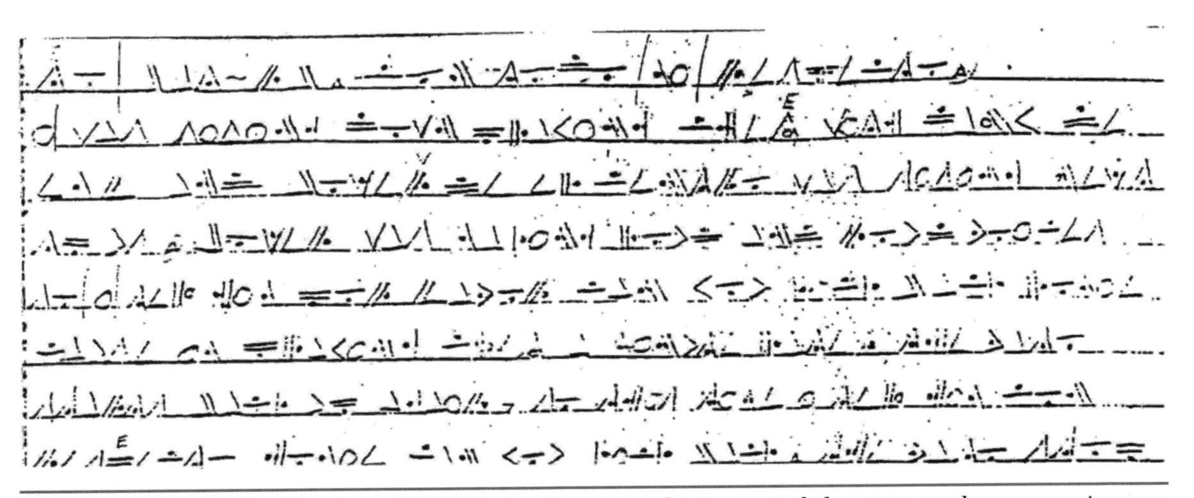

**Figure 4.** *Another application of MCMC. A prison guard intercepted the message between prisoners, written in an unknown code. The message was deciphered using MCMC, assuming that each symbol represents a different letter and that the likelihood of two letters being next to each other is similar to English. Thus the digraph "QA" should be extremely rare, while the digraph "TH" should be quite common. Decryption was successful even though the message contained significant numbers of Spanish words and prison slang. (Figure courtesy of Persi Diaconis from the course work of Marc Coram.)*

In the example of the prisoner's message, Coram had a handy objective measure of when convergence had occurred—namely, the message made sense. In fact, he did not even have to run the Monte Carlo simulation more than once. The probability distribution had a sharp peak at the one correct configuration, and one run of the algorithm was enough to detect that peak.

In general, though, there is no obvious criterion for stopping. If you stop too soon, your simulation will not have enough time to move away from your initial guess, $\sigma_0$. If you wait too long, then you use up computing time for no good reason.

Several ways have been proposed to get around this difficulty. One method, called coupling from the past (CFTP), diagnoses all by itself when to stop, and produces "perfect" samples from the target probability distribution. (By contrast, Metropolis' algorithm produces samples that have probabilities *close* to the target probabilities.)

The idea of CFTP, which was proposed in 1996 by Jim Propp and David Wilson, is that the output configuration of an arbitrarily long run of the MCMC algorithm can often be inferred *without* knowing the initial guess $\sigma_0$. In the case of the Ising

**The trouble with CFTP is that you can't predict in advance how long you will have to wait for your perfect, unbiased sample—a thousand steps or a million! In practice, though, its run times seem comparable to standard MCMC algorithms.**

model, here is how CFTP works. To run the Metropolis algorithm, you must accept or reject a proposed change of the spin at one lattice point, with a probability equal to $\mu(\sigma_1)/\mu(\sigma_0)$. In practice, you do this by using your computer to generate a random number $a_1$ between 0 and 1. If $a_1 \leq \mu(\sigma_1)/\mu(\sigma_0)$, you accept the change; otherwise you reject it.

Now Propp and Wilson exploited a previously known monotonicity property of the Ising model. Suppose you start with two different starting configurations ($\sigma_0$ and $\sigma_0'$) such that $\sigma_0$ has a $+1$ in every position, and $\sigma_0'$ has a $-1$ in every position. Or, more generally, suppose that $\sigma_0$ dominates $\sigma_0'$, in the sense that it never has a $-1$ spin where $\sigma_0'$ has a $+1$. If you apply the Metropolis algorithm *with the same random number seeds* $a_1, a_2, a_3, \ldots$ to these two configurations, the dominant one will remain dominant, but the difference between their total magnetizations can never go up—at every step, it decreases or remains the same. Eventually, the difference will decrease to zero and the two configurations will coalesce. In fact, *all* initial configurations will "coalesce" into the same final configuration. Thus the dependence on the initial guess $\sigma_0$ is removed, and the coalesced configuration is an unbiased sample from the target probability distribution. Not only that, it is easy to tell when coalescence has occurred.

CFTP seemed almost like magic at the time. Unfortunately, there is no free lunch. In CFTP you still can't predict in advance how long you will have to wait for your perfect, unbiased sample—a thousand steps or a million! In practice, though, its run times seem comparable to standard MCMC algorithms.

A somewhat more principled approach is to estimate the burn-in time in advance. That is where the cutoff phenomenon, and the work of Lubetzky and Sly, enters the picture.

There is one classical constraint on the mixing time. If the system has $n$ states, then the Markov chain is completely described by an $n$-by-$n$ matrix of transition probabilities from each state to each other state, as illustrated in Figure 2, page 61. Running the chain for many steps corresponds to taking powers of this matrix. Unlike powers of a number, the powers of the matrix settle down rather quickly to a constant, and the (identical) rows of this constant matrix give the stationary distribution. A number called the *spectral gap*, which can be computed algebraically, estimates how long it takes for the tardiest row to "settle down." This time is called the "relaxation time" of the Markov chain.

It may seem surprising at first that the relaxation time is not necessarily the same as the mixing time or the burn-in time. Here is a simple analogy: Consider $n$ coins in a row of $n$ boxes. All of the coins start heads up. On each turn you select a box at random, and then flip the coin in that box. How many turns would you expect to take before the orientation of the coin in box one is randomized? Because it only takes one flip to randomize the coin, this boils down to asking how long it will be before you select box one. Because each box is equally likely, you would expect, on average, to need $n$ turns. That is the relaxation time for the system, the time it takes to "forget" the initial orientation of *one* coin.

However, the mixing time is the number of turns it takes to forget the whole starting configuration, or to randomize the orientation of *every* coin. It's quite unlikely (especially if $n$ is large) that you will select each box exactly once in the first $n$ steps. For that reason, you would expect the mixing time to be larger than $n$. In fact, the expected number of turns it takes to flip every coin is $n \log n$.

Actually, Lubetzky says, there is a very subtle reason that it only takes *half* that long to randomize the orientation of the coins. The coins only need to be random enough to pass any statistical test. When you flip $n$ coins, the number of heads you get is usually *not* equal to $\frac{1}{2}n$. You are likely to have some small deviation from that number, and the size of the deviation is typically $\sqrt{n}$. So even if you leave $\sqrt{n}$ coins *unflipped*, and therefore have $\sqrt{n}$ *extra* heads, there is no statistical test that will be able to tell the excess apart from chance variation. For that reason, the mixing time in this case is actually the amount of time it takes (on average) to select all but $\sqrt{n}$ of the boxes, and it can be shown that the expected number of turns is $\frac{1}{2}n \ln n$. However, even allowing for this subtlety, the most important thing to observe is that the mixing time ($\frac{1}{2}n \ln n$) is significantly greater than the relaxation time ($n$).

**A somewhat more principled approach is to estimate the burn-in time in advance. That is where the cutoff phenomenon, and the work of Lubetzky and Sly, enters the picture.**

This example, called the hypercube random walk, is very illustrative in spite of its simplicity. It shows that abrupt mixing is a *statistical* phenomenon, a result of a lot of small-scale events that happen around the same time. "We want to understand the process of a system that starts out of equilibrium, and ask how it reaches equilibrium," says Peres. "There are two main ways it can do this. First, you can wait for one rare event to become likely. The low-temperature Ising model is the example of that. For the system to reach equilibrium, you have to wait long enough for it to switch from a mostly-plus to mostly-minus configuration. There is no cutoff in this case, because the waiting time has an exponential distribution."

"Another way to reach equilibrium is that your system has a lot of independent components, and you need almost all of them to forget their starting state. That is what happens in the hypercube model," Peres says. And, he points out, "The random walk on the hypercube is just an infinite-temperature Ising model."

The fact that mixing in the Ising model occurs abruptly at infinite temperatures suggested that the same thing should hold at any "high" temperature—that is, any temperature above the phase transition. However, an analogy is not the same as a proof. In 2007, Peres and two colleagues proved the cutoff phenomenon for the Ising model on a *complete graph*, in other words a crystal where every atom touches every other one. However, this is clearly not physically realistic. "The lattice model was considered the prime motivation and the big prize, but I considered it to be a prize that was a long way off," Peres says.

Then, in 2008, Lubetzky and Sly began working on the problem—Lubetzky as a researcher at Microsoft Research and Sly as a summer intern visiting from the University of California at Berkeley, where he was a graduate student. Together they made spectacular progress. "At the end of August we announced to our friends that Peres' conjecture for the Ising model was settled," Lubetzky says. "But it took us a year longer to write it down."

"Usually, to solve a good open problem, you need a good idea, sometimes two," Lubetzky says. "In this paper we had to introduce three new things." Probably the most important of the three new ideas was the simplest one. Lubetzky and Sly asked themselves which vertices in the initial configuration actually matter when the lattice reaches its mixing time. The answer is that very few of them do. (See **Fading Memory**, page 76.) Most of them get overwritten long before the information in the initial configuration can propagate through the rest of the lattice.

Their idea was quite similar to Propp and Wilson's coupling from the past, where the Markov chain is allowed to run so long that *none* of the vertices in the original configuration matter any more. In Lubetzky and Sly's proof, the chain does not run that long—but it runs long enough that the vertices that matter are few and far between. Hence the high-temperature Ising model starts to resemble the infinite-temperature hypercube model. The sites that matter evolve independently of each other, and the other sites ... don't matter.

**"Usually, to solve a good open problem, you need a good idea, sometimes two," Lubetzky says. "In this paper we had to introduce three new things."**

Of course there were difficulties involved in sorting out what phrases like "almost independent" and "few and far between" mean exactly, and showing that they were sufficient to imply the cutoff phenomenon. That was why the details took a year to work out. Eventually, Lubetzky and Sly computed the cutoff time precisely—even in three or more dimensions, where the critical temperature of the phase transition is still unknown. The mixing time for the MCMC algorithm on the Ising model is $T = \frac{d}{2\log(\lambda_\infty)} n \log n$, where $d$ is the dimension of the lattice, $n$ as before is the number of vertices, and $\lambda_\infty$ is the spectral gap for the "infinity-by-infinity" lattice. (See Figure 5)

The methods should apply to many other MCMC models. The minimum requirements seem to be that CFTP should work, and that the MCMC algorithm should be defined on a *family* of graphs (such as the family of $\sqrt{n}$-by-$\sqrt{n}$ lattices), so that it makes sense to talk about what happens as $n$ goes to infinity. For instance, Lubetzky and Sly have shown that the cutoff phenomenon occurs at *any* temperature in the 2D Ising model if there is an external magnetic field, and they have also shown it occurs at sufficiently high temperatures in the Potts model, where the two spins $+1$ and $-1$ are replaced by $q$ different colors that can be placed at any vertex.

The late Alfréd Rényi joked that "a mathematician is a device for turning coffee into theorems" (a quote also attributed to his colleague Paul Erdős). But in the case of abrupt mixing, they were clearly wrong: it takes two mathematicians, plus milk.

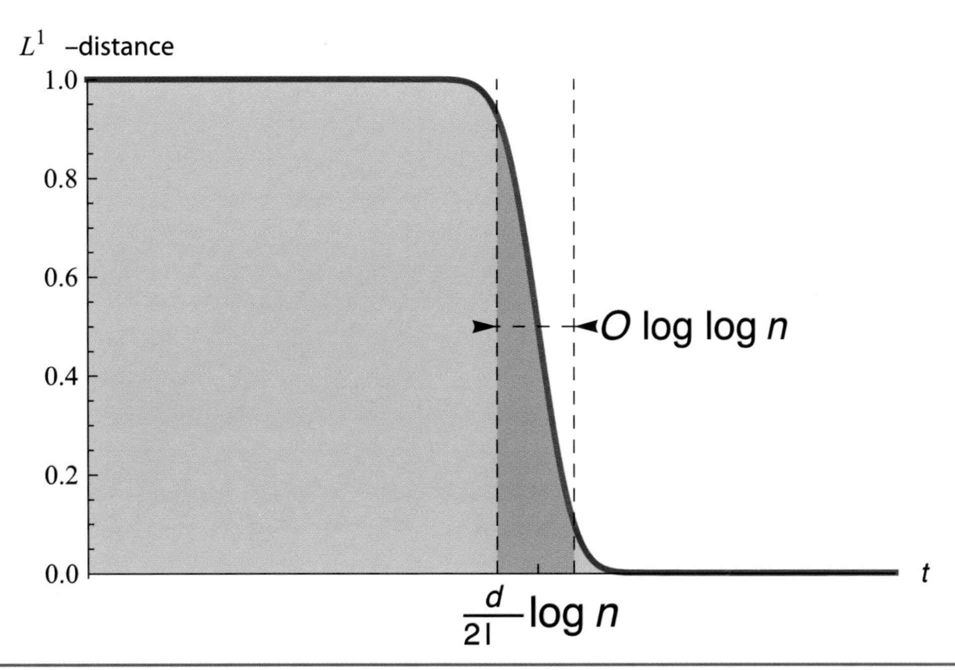


**Figure 5.** *Cutoff phenomenon in MCMC applied to the Ising model. If the number of iterations of the MCMC method is too small, the sample produced will be far from random, and strongly biased by the initial configuration. Within a very short time window (purple region) the sample distribution moves rapidly toward the equilibrium probability distribution. Past this window, further iterations of the MCMC are essentially useless because the sample distribution and equilibrium distribution are almost identical. Mixing is not only rapid, but also very abrupt. (Figure courtesy of Eyal Lubetzky and A. Sly.)*

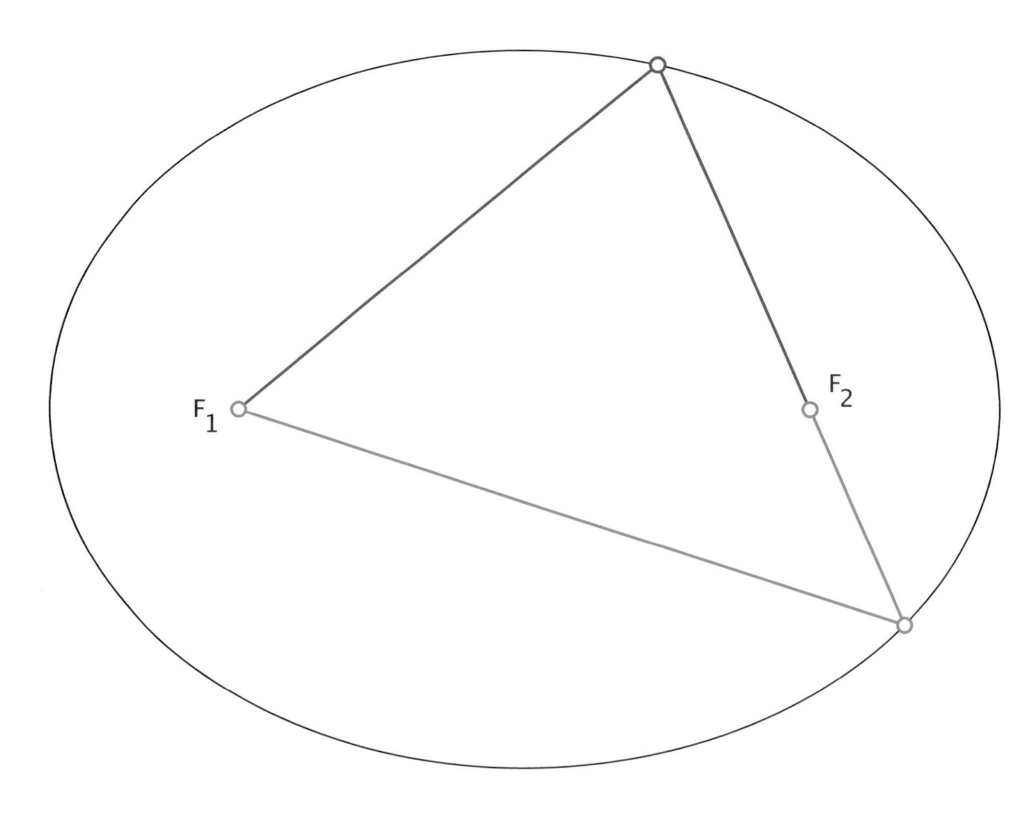


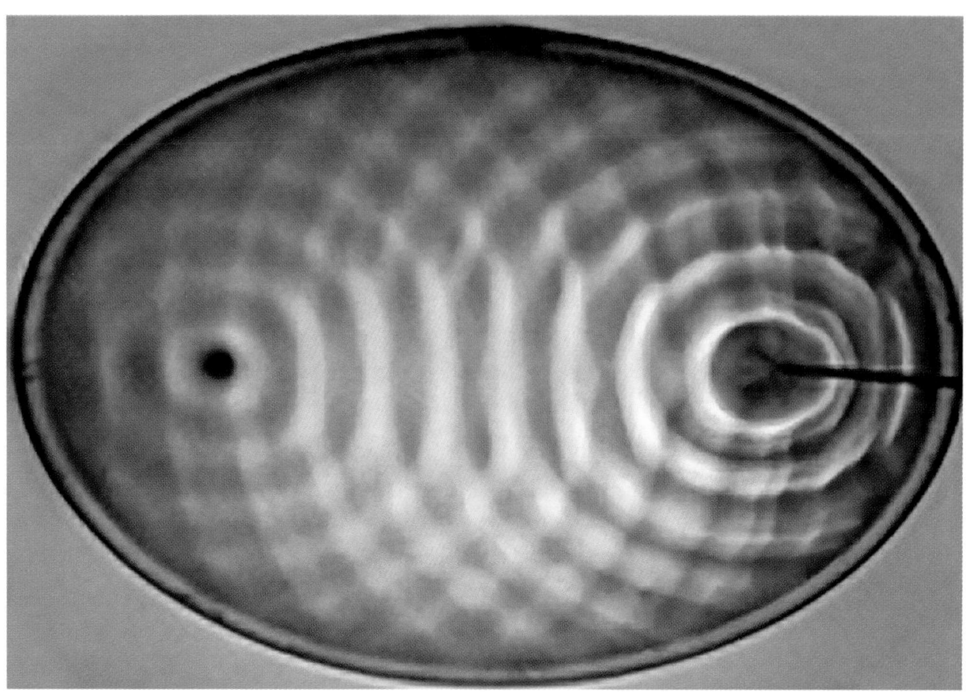

**Figure 1.** *(Top) Billiard-ball trajectories (red and blue) that start at one focus of an ellipse will always pass through the other focus. (Bottom) In a "wave" version of billiards, waves that are generated at one focus of an elliptical tank tend to concentrate at the other focus. (The bottom figure is courtesy of Alex Helman, Physics Department, University of California at Santa Cruz.)*

# In Search of Quantum Chaos

When chaos theory arrived on the mathematical scene in the 1970s and 1980s, it revolutionized the study of classical dynamical systems. Mathematicians suddenly realized that many apparently deterministic processes are inherently unpredictable, because even the slightest change in the initial data could completely change the long-term behavior of the system. For example, the flapping of a butterfly's wings in Australia today could cause a hurricane in the Caribbean, two weeks later.

Chaos can be found even in much simpler systems than the weather. For example, the motion of a billiard ball on a frictionless billiard table with at least some non-straight sides (think billiards on an ice hockey rink) is usually chaotic. There is essentially no way to predict where the ball will be after it has hit the bumpers a few hundred times.

Although chaos seemed rampant in macroscopic systems, it seems to be absent in the atomic and subatomic realm. (Perhaps this is good; we would not want the atoms in our bodies to start flying apart unpredictably.) One reason is that quantum mechanics is built on *linear* differential equations, such as the Schrödinger equation. But chaos only arises in systems with *nonlinear* feedback, a sort of cascade of events that magnifies the "want of a nail" into the "loss of a kingdom."

However, an interesting gray zone exists between the quantum world and the macroscopic world—the world of the "semiclassical limit." Quantum physics says that the world is grainy; energy comes in discrete chunks. But if you let the grain size get smaller—or, equivalently, if you look at systems with more and more energy—the quantum "graininess" should gradually become invisible. The world should look more and more like the world of classical physics. In short, you should start seeing chaotic behavior.

Mathematicians have recently confirmed the occurrence of "quantum chaos" in this gray zone. Roman Holowinsky of The Ohio State University and Kannan Soundararajan of Stanford University (who goes by his last name, or "Sound" for short) proved a version of the Quantum Unique Ergodicity (QUE) conjecture, which says that a semiclassical quantum billiard ball is just as unpredictable as a regular billiard ball.

Unfortunately, the shapes of the billiard tables that Holowinsky and Soundararajan's argument applies to are very specific, so the physical applications of their theorem may be limited. But in a field starved for mathematically tractable examples, any rigorous proof of chaos is a big step forward. From a mathematical point of view, it is equally noteworthy that the study of this seemingly applied problem has led to new discoveries in the pure realm of number theory.

**Roman Holowinsky.** *(Photo copyright C. J. Mozzochi, Princeton, NJ.)*

**Kannan Soundararajan.** *(Photo copyright C. J. Mozzochi, Princeton, NJ.)*

## From Ellipses to Negatively-Curved Manifolds

To understand what the Quantum Unique Ergodicity conjecture says, let's start with a visit to the Oval Office. In June 2010, a group of junior high-school students who had won the Raytheon MATHCOUNTS national competition visited President Barack Obama in his famous elliptical office and surprised him with the question, "Where are the foci?"

According to news reports, Obama was intrigued enough (after the students explained to him what a focus was) that he joined them in pacing out the length and breadth of the room and estimating the location of the two focal points. While Obama's aides nervously checked their watches, the middle-schoolers and the president eventually answered what the *San Jose Mercury News* called "just about the coolest math question of all time"—the foci are about 10.5 feet from the center of the office.

What is so cool about the focus of an ellipse? If you put a billiard ball at one focus and hit it in *any* direction, it will bounce off the wall and pass through the other focus. (We are ignoring the effects of spin.) In other words, on an elliptical billiard table the ball's motion is exactly the opposite of chaotic: It is totally predictable. You can even predict the time when the ball will arrive at the other focus, because the length of the bouncing path is independent of the direction. (See Figure 1, page 88, top.)

The same phenomenon holds if you replace billiard balls by sound waves. If you stand at one focus and whisper, a person who is standing at the other focus will be able to hear distinctly what you are saying. Intuitively, that is because the sound waves produced by your vocal cords will go out in all directions, bounce off the walls (just like the billiard ball) and then congregate again at the other focal point. (See Figure 1, page 88, bottom.)

This rough explanation, however, belies an increase in mathematical subtlety. Sound waves do not travel in rays, as little "balls" of sound. They are more like the vibrations of a drumhead, spreading out in all directions. If $\varphi(x, y)$ represents the amplitude of the vibrations at any point $(x, y)$, then $\varphi$ satisfies a differential equation that looks like this:

$$\Delta\varphi + \lambda\varphi = 0.$$

Here $\Delta = \frac{\partial^2}{\partial x^2} + \frac{\partial^2}{\partial y^2}$ is the Laplace operator, and $\lambda$ is a constant related to the frequency of the wave. The equation will have solutions for some values of $\lambda$ but not for others. The values that do have solutions correspond to different "modes of vibration" of the drumhead. (See Figure 2)

The process of quantizing a classical problem follows a similar procedure. In quantum physics, all particles are waves, represented by a "wave function" $\psi(x, y, t)$. The values of $\psi$ are complex numbers, and the amplitude $|\psi(x, y, t)|^2$ represents the probability of observing the particle at position $(x, y)$ at time $t$. The waves satisfy a version of the wave equation known as Schrödinger's equation. Of particular interest are the standing wave solutions, which take the form $\psi(x, y, t) = \varphi(x, y)e^{2\pi i\lambda t}$, where $\varphi(x, y)$ is a solution to the same equation as above. For such solutions, the probability of observing the particle at position $(x, y)$ does not change over time.

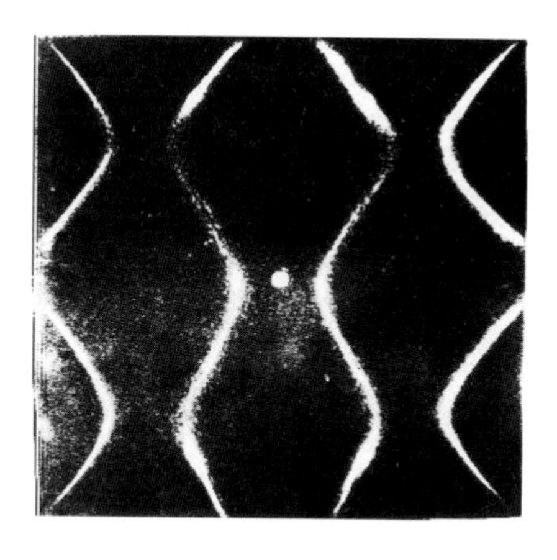
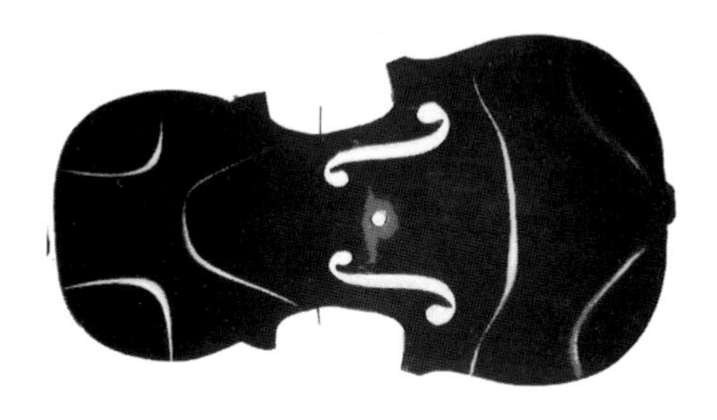

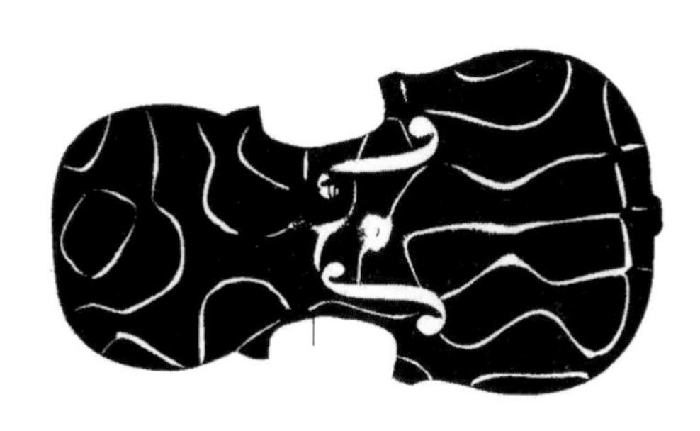

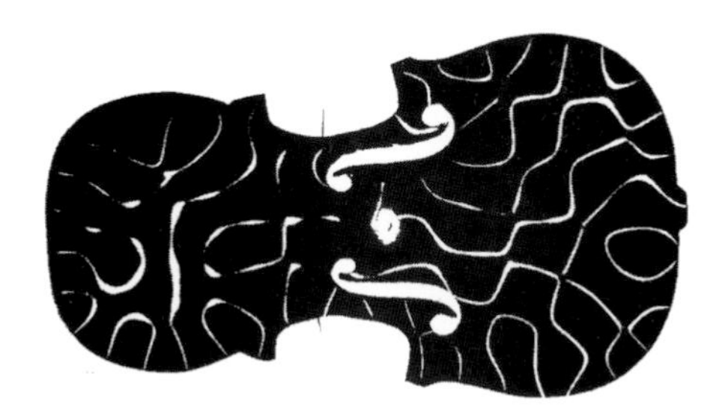

**Figure 2.** *Chladni patterns have been used for two centuries to visualize acoustic vibrations. Sand grains concentrate on the "nodes", where the amplitude of vibration is close to zero. Each mode of vibration has its own unique frequency and its own distinctive Chladni pattern. Left: The vibrations of a square plate at frequencies of 142, 1450, and 3678 hertz; Right: the vibrations of a violin-shaped plate at frequencies of 145, 762, and 1452 hertz. (Figures courtesy of Stephen W. Morris, Nonlinear Physics, University of Toronto.)*

**In 1974, Alexander Shnirelman proved the Quantum Ergodicity Theorem, which implies that as $\lambda$ increases, *almost all* of the modes of vibration will have probability distributions that get closer and closer to a uniform gray. In this case, mathematicians say that the semiclassical billiard ball is *ergodic* (and that is the reason for the term, "quantum ergodicity").**

The values of $\lambda$ for which solutions exist, denoted $\lambda_1, \lambda_2, \lambda_3, \ldots$, are called eigenvalues of the Laplace operator, which represent the energy of the wave. Thus, if we are interested in the semiclassical limit, we should ask what happens to the solutions for $\varphi$ as $\lambda$ approaches infinity.

Figure 3 shows the modes of vibration $\varphi(x, y)$ for quantum billiards on three differently shaped tables. In each case, we are looking at very high-energy modes that are far out on the sequence of eigenvalues, around $\lambda_{5600}$. In all of the pictures, the quantum billiard ball focuses in the dark regions, which means that those regions are the most likely places to observe the quantum billiard ball at any given time. If the motion of the quantum billiard ball were completely unpredictable, you would see a uniform expanse of gray.

In Figure 3a, that is not what you see at all. In fact, many of the quantum wave functions hug the outside of the ellipse and never even enter the central region. This corresponds to billiard balls that stay close to the edge of the table, or sound waves that stay close to the walls of the room. That is precisely analogous to the motion of a non-quantum, classical billiard ball on an elliptic table. No chaos here.

Contrast this with the situation in Figures 3b and 3c, known as the "Bunimovich stadium" and the "Barnett billiard table" respectively. In both of these cases, a classical billiard ball will have chaotic orbits, so that the likelihood of seeing it at any particular point on the table is the same. Thus, in the semiclassical limit, we would expect the amplitude function to approach a uniform gray. Is that what happens?

In the case of the stadium, the answer is yes and no. The amplitudes tend to cluster around periodic (closed) trajectories of a classical billiard ball. These are unstable in the classical case, but the quantum billiard still finds them. For example, in the bottom row of Figure 3b there is one mode where the amplitude concentrates around an especially simple billiard path, where the ball bounces vertically up and down forever. On the other hand, most of the amplitudes do not seem to be as focused as the amplitudes for the ellipse problem.

Both of these empirical observations can be proved. In 1974, Alexander Shnirelman proved the Quantum Ergodicity Theorem, which implies that as $\lambda$ increases, *almost all* of the modes of vibration will have probability distributions that get closer and closer to a uniform gray. In this case, mathematicians say that the semiclassical billiard ball is *ergodic* (and that is the reason for the term, "quantum ergodicity").

On the other hand, in 2008 Andrew Hassell proved that the "bouncing ball" modes for the Bunimovich stadium do not go away as $\lambda$ approaches infinity. No matter how high the energy is, you can still find modes of vibration whose amplitudes are not distributed uniformly. This phenomenon is called "scarring," because the dark gray regions resemble a scar.

Finally, in Figure 3c, the Barnett billiard table, there is no apparent scarring. It is still true that the individual amplitudes are far from uniform. Nevertheless, from the numerical evidence it appears that the distribution gets closer and closer to uniform gray as $\lambda$ increases, so that the semiclassical billiard ball is ergodic. Note, in particular, that "whispering gallery" modes, as in the case of the ellipse, do not exist, because the outward-curving walls disperse the waves rather than focusing them.

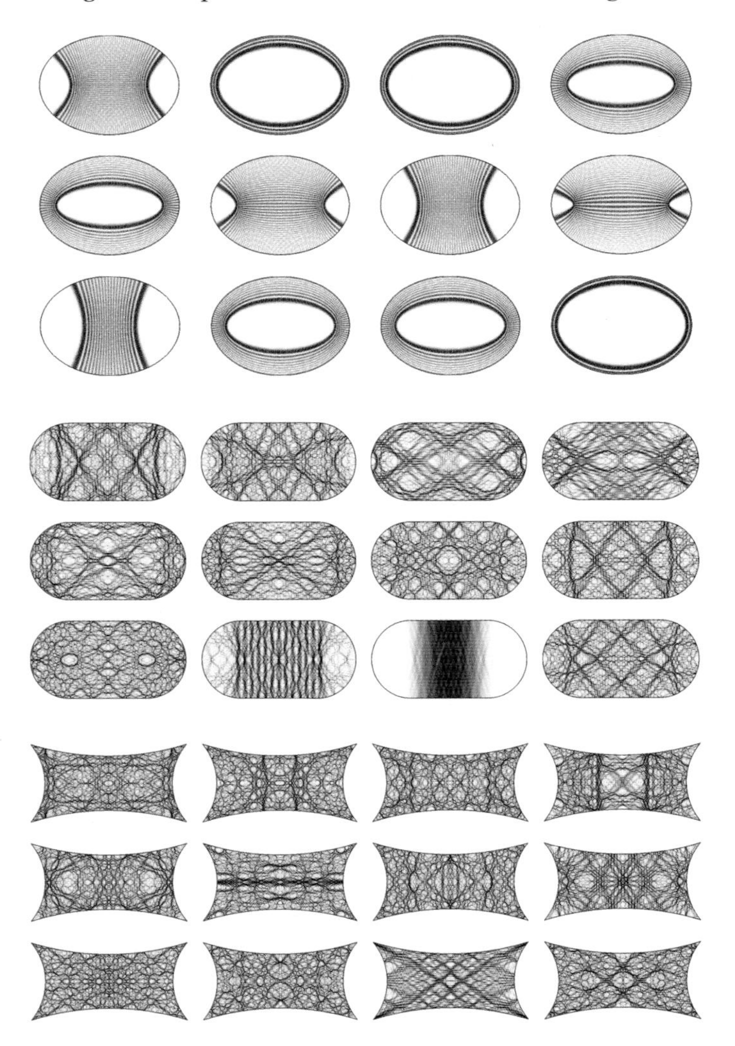

**Figure 3.** *Computed "Chladni patterns" for a quantum billiard ball on (top 3 rows) an elliptical table; (middle 3 rows) a stadium-shaped table; (bottom 3 rows) a table with a "dispersing" boundary. The darkness at any location corresponds to the probability of finding the quantum billiard ball there. All of the modes in the top 3 rows and some of the modes in the middle 3 rows exhibit scarring—a strong tendency for the ball to frequent certain places and avoid others. (Figures courtesy of Alex Barnett)*

On the basis of examples like these, Peter Sarnak of Princeton University and Zeev Rudnick, who is now at Tel Aviv University, formulated the Quantum Unique Ergodicity conjecture in the early 1990s. For convenience they phrased the conjecture in terms of geodesic (straight-line) motion on negatively curved manifolds, which are closed surfaces without boundaries. However, the QUE conjecture can be adapted also to examples like Barnett's, where there are concave boundaries that cause nearby billiard trajectories to disperse. However, the QUE conjecture does *not* apply to examples like 1a and 1b,

**Zeev Rudnick.** *(Photo copyright C. J. Mozzochi, Princeton, NJ.)*

where the walls are convex or straight. Sarnak and Rudnick hypothesized that in the case of negative curvature (or dispersing boundaries) not only do *most* modes of the quantum billiard become ergodic in the semiclassical limit, as Shnirelman had proved, but in fact they *all* do. The uniform gray is the *unique* limiting distribution, or to put it differently, there is no scarring. If the conjecture is true, then examples 3a (non-ergodic), 3b (ergodic but not uniquely ergodic) and 3c (uniquely ergodic) would all be quite distinct from one another. In some sense, example 3c would be the most chaotic.

## The Madras Connection

In the early 1900s, Srinivasa Ramanujan, a self-taught clerk from Madras (now Chennai), India, discovered hundreds of beautiful and sometimes bizarre mathematical equations, most of which had previously been unknown. Rather frustratingly, he seldom gave even a semblance of a proof. Ramanujan himself attributed his formulas to the goddess Namagiri. Because mathematicians do not accept divine fiat as a justification for mathematical theorems, it has fallen to later generations to prove (or, in very rare cases, refute) Ramanujan's results. Ramanujan himself died in 1920, at the age of 32.

As a teenager growing up in Ramanujan's home city, Soundararajan knew all about Ramanujan, who is now a national hero in India. But he could hardly have suspected that he would one day make use of one of Ramanujan's most famous conjectures—and extend it—in order to understand quantum chaos. "The problem Holowinsky and I solved uses Ramanujan's bound for the Fourier coefficients of holomorphic modular cusp forms," Sound explains.

Holomorphic modular forms are complex-valued functions of a complex number $z = x + iy$ in the half-plane $y > 0$, which have a unique property of being nearly invariant under a particular kind of coordinate change. If we define a new coordinate by the equation $z' = \frac{az+b}{cz+d}$ (where $a, b, c$, and $d$ are ordinary integers with $ad - bc = 1$), then $f(z') = (cz + d)^k f(z)$ for some integer $k$, called the "weight" of the modular form. Of course it would be simpler if $f(z') = f(z)$, making $f$ completely invariant under the coordinate change. However, holomorphic functions must satisfy an extra constraint, the Cauchy-Riemann equations, which can be written compactly as $\left(\frac{\partial}{\partial x} + i\frac{\partial}{\partial y}\right) f = 0$. If we want $f$ to be holomorphic, then we have to settle for this near-invariance, which goes under the name of "modularity."

Modular forms are exceedingly rare and important for number theory, and thus they have been very thoroughly studied over the years. It is known that the number of independent modular forms of weight $k$ is roughly $k/12$, and thus the first one has weight 12. It is called the discriminant modular form, denoted by $\Delta$. This function attracted Ramanujan's attention because of its beautiful number-theoretic properties. For instance, the coefficients are unexpectedly connected to the number of ways to write a number as a sum of 24 perfect squares.

Although modular forms are not quite invariant under general coordinate transformations of the form $z' = \frac{az+b}{cz+d}$, they are invariant when $c = 0$ and $d = 1$; thus, $\Delta(z) = \Delta(z+1)$. This makes $\Delta$ a periodic function. Hence it can be expanded as a Fourier series:

$$\Delta(z) = e^{2\pi iz} - 24e^{4\pi iz} + 252e^{6\pi iz} + \ldots$$

The coefficients $1, -24, 252, \ldots$ are called the values of the tau function $\tau(n)$. Ramanujan conjectured that the values of the tau function are built up in a simple way from its values at the prime numbers. For example, to get $\tau(6)$, we simply multiply $\tau(2)$ and $\tau(3)$: $\tau(6) = -24 \times 252 = -6048$.

Ramanujan also conjectured that the prime values $\tau(p)$ satisfy a simple inequality: $|\tau(p)| \le 2p^{11/2}$. Although Ramanujan left hundreds of formulas unproved in his lifetime, this is the only one that actually became known as "The Ramanujan Conjecture." It was finally proved in 1974 by the Belgian mathematician Pierre Deligne, who won a Fields Medal for his work (and was also made a viscount of Belgium).

As is often the case, the solution to a good mathematical problem reveals a lot of hidden structure and leads to much more general results. (This is one good reason not to simply take Namagiri's word!) Deligne actually proved that Ramanujan's conjecture was true for any modular form that satisfies an extra symmetry property, which can be roughly stated as follows: the value of $f(z)$ is proportional to a weighted average of $f$ at a special set of $p+1$ points surrounding $z$. The "averaging" operator is called the *Hecke operator*, and modular forms with the extra symmetry are *Hecke eigenforms*. Hecke had proved that any modular form can be built up out of Hecke eigenforms of the same weight. Since there is only one modular form of weight 12—namely, Ramanujan's $\Delta$—it must be a Hecke eigenform, and therefore Ramanujan's inequality holds for it.

**Peter Sarnak.** *(Photo copyright C. J. Mozzochi, Princeton, NJ.)*

## Belt or Suspenders?

Now it's time to bring the quantum wave functions back into the story. As a result of the Cauchy-Riemann equations, modular forms $f(x+iy)$ satisfy the equation $\Delta f = 0$, which is extremely reminiscent of the equation that constrains the amplitude of a quantum wave function, $\Delta\varphi = \lambda\varphi$. The former equation lacks the energy quantization that is provided by $\lambda$. But modular forms are quantized in a different way, by the weight of the modular form, $k$. In some ways that is more convenient, because $k$ only takes integer values, while the sequence of eigenvalues $\lambda_1, \lambda_2, \lambda_3, \ldots$ is somewhat irregular and not yet completely understood. (For example, no one knows yet whether two different modes of vibration can share the same eigenvalue $\lambda$.)

Rudnick and Sarnak believed that modular forms would behave very much like quantum wave functions defined on the region shown in Figure 4, which is called the fundamental domain of the modular group. The fundamental domain acts like a single room in a hall of mirrors, whose multiple reflections fill out an entire plane (or, in this case, a half-plane). Any point $z'$ in the half-plane is the "mirror image" of some point $z$ in the fundamental domain, i.e., $z' = \frac{az+b}{cz+d}$, and thus the value of $f(z')$

anywhere in the half-plane is determined by the value of $f(z)$ in the fundamental domain.

Unfortunately, the most obvious version of QUE for modular forms turns out to be false. If $\Delta(z)$ is the discriminant modular form, then $\Delta(z)^n$ is also a modular form of weight $12n$. It will have zero amplitude exactly where $\Delta(z)$ does, and maximum amplitude exactly where $\Delta(z)$ does. Thus, as $n$ approaches infinity, the amplitudes of $\Delta(z)^n$ do *not* fade to a uniform gray.

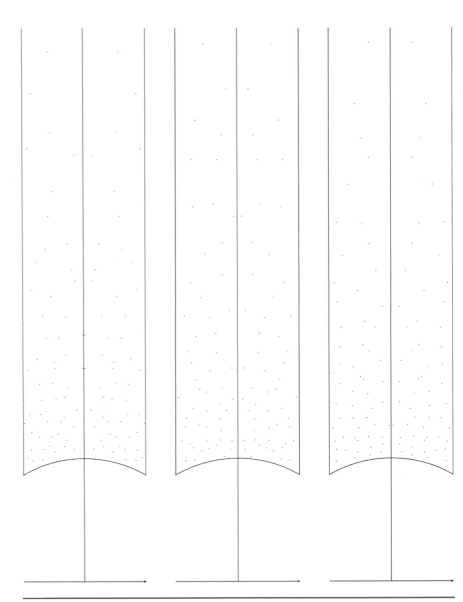

**Figure 4.** *Computed "Chladni patterns" for Hecke eigenforms. Note that these look quite different from Figures 2 and 3, because a Hecke eigenform (a complex-valued function) takes the value zero only at isolated points (blue dots). The principle of unique ergodicity implies that as the dots become more numerous, they should not form scars. They should only show a gradual decrease in density with height above the x-axis. (Figures courtesy of Fredrik Strömberg.)*

However, Rudnick and Sarnak conjectured that QUE *does* hold if you only look at Hecke eigenforms. Note that $\Delta(z)$ is a Hecke eigenform, but $\Delta(z)^n$ is not, so this "holomorphic QUE conjecture" does not run afoul of the easy counterexample mentioned in the previous paragraph.

By 2008, Soundararajan, a former student of Sarnak, was getting close to proving the holomorphic QUE conjecture. Another former student of Sarnak's, Tom Watson, had discovered a beautiful way to express the level of grayness in terms of an L-function on the line $x = \frac{1}{2}$. The term "L-function" might be unfamiliar to most people, but for number theorists L-functions are pure gold. They are a general class of functions with three nice properties:

1. They can be written as an infinite sum of the form $L(s) = \frac{a_1}{1^s} + \frac{a_2}{2^s} + \frac{a_3}{3^s} + \dots$, defined for any complex number $s = x + iy$ (usually with $x > 1$).
2. They satisfy a symmetry property called a functional equation, which allows them to be defined for $x \leq 1$, even when the infinite sum does not converge.
3. The line of symmetry is usually the line $x = \frac{1}{2}$ (or $s = \frac{1}{2} + iy$).

Almost any sufficiently deep question about the overall distribution of prime numbers usually comes down to a question about the behavior of an L-function inside the critical strip, $0 \leq x \leq 1$, and particularly on the line of symmetry, $x = \frac{1}{2}$. (The prototypical example is the Riemann Hypothesis, which deals with the simplest L-function, which is called the Riemann zeta function.)

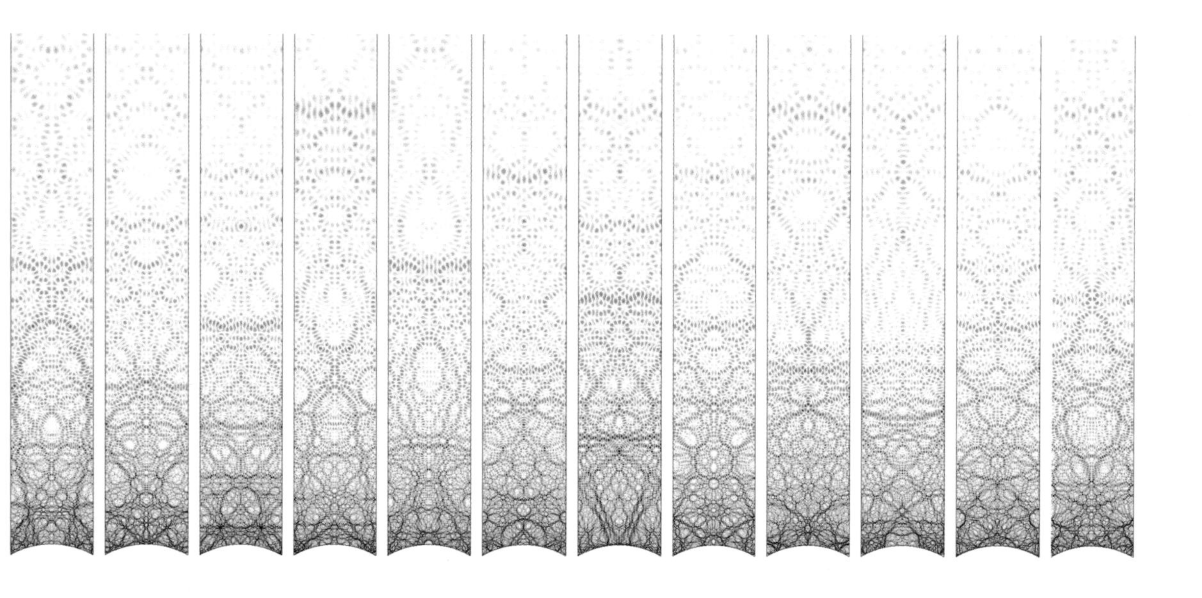

**Figure 5.** *Computed "Chladni patterns" for the quantum billiard ball on the same region as Figure 4. Quantum unique ergodicity (QUE) implies that at very high energies, the density of gray should be inversely proportional to height above the x-axis (the bar at the bottom), with no scarring. Unique ergodicity has now been proved both for Figure 4 and Figure 5. (Figure courtesy of Alex Barnett.)*

**Number theorists have a curious sort of love-hate relationship with the generalized Riemann Hypothesis. Virtually everyone thinks it is true. It represents what should be true in the "best possible world,"...**

The unique ergodicity of Hecke eigenforms does not, at first blush, look like a question about prime numbers, but it is. Remember that those are precisely the modular forms whose prime-number coefficients $a_p$ were proved, by Deligne, to satisfy a Ramanujan inequality. Sound's strategy, following Watson, was to package those coefficients up into an L-function and see what happened.

It turned out that the L-function needed to grow relatively slowly on the line of symmetry. In fact, there is a generalized version of the Riemann Hypothesis that would imply that the L-function grows *much* slower than needed, and so the unique ergodicity would follow very easily.

Number theorists have a curious sort of love-hate relationship with the generalized Riemann Hypothesis. Virtually everyone thinks it is true. It represents what should be true in the "best possible world," if the goddess Namagiri is smiling upon us. Many theorems have been proven using the generalized Riemann hypothesis as an assumption. However, such theorems can only be considered to be provisionally true. So it becomes very important to try to prove them independently of the GRH—both so that we can know they are unconditionally true, but also to gain some insight into the workings of the GRH itself. "Every physicist who works on quantum chaos has a secret dream to understand the Riemann hypothesis," Rudnick says.

The key to proving QUE without the Riemann hypothesis, Sound realized, was to milk a little bit more information out of Ramanujan's conjecture. Deligne had proved that the coefficients $a_p/p^{(k-1)/2}$ were bounded. Sound went one step farther and proved that the *average* value of a related series of coefficients actually decreases, very slowly, as $k$ increases. In other words, some cancellation occurs. This was enough to give him the bound he needed for the L-function.

But there was one problem. For a small (and perhaps only hypothetical) set of Hecke eigenforms $f$, Sound's argument did not work. That was where matters stood in the spring of 2008, when his colleagues Philippe Michel and Akshay Venkatesh told him that Holowinsky was working on the same problem. Holowinsky, it turned out, had run into the same problem—he could prove the QUE for all but a handful of Hecke eigenforms. Sound invited Holowinsky to Stanford to compare notes, and they discovered that a small miracle had occurred: Their work was completely complementary. The few forms that defied Sound's proof would yield to Holowinsky's, and vice versa. Choosing between the proofs was like choosing between a belt or suspenders. For most people, either one will work; but for a few people, only one will do the job. By September, the two mathematicians had finished their "belt-and-suspenders" proof.

At this point, they had proved only the holomorphic version of QUE (see Figure 4, page 86), but not the "real" version for quantum waves (see Figure 5, page 87). However, they had broken a psychological barrier.

Several years earlier, Elon Lindenstrauss had nearly proven QUE for the regions in Figures 4 and 5; however, he could not rule out a sequence of modes of vibration that would let the billiard ball escape out the open top of the region. In this case, the semiclassical limit would not be a uniform gray—it would be a uniform white, corresponding to no billiard ball at all. (Another bizarre possibility was a uniform light gray, where "part" of the billiard ball would escape to infinity.) It often happens that solving one problem will show you how to get past the sticking point in a related problem. Sound realized that the only thing missing was an analogue of the Ramanujan conjecture, which had worked so well for the holomorphic case. It took him just one day during winter vacation in 2008 to find the necessary argument to complete Lindenstrauss's proof. "There is no rogue mass escaping out to infinity," he says.

The whole story is a beautiful example of the interplay between physics and mathematics. The physical problem motivated Sarnak and Rudnick to propose a number-theoretic analogue that might never have occurred to them without the stimulus from quantum chaos. In that setting, extra tools were available—the Ramanujan inequality, the whole machinery of L-functions and Hecke eigenforms—that are not available in general. However, once the path to the solution became clear in the holomorphic case, it cracked open a psychological barrier that had been stopping progress in the physical case.

At this point a full proof of QUE is still, in all likelihood, very far away. The strategy employed by Soundararajan and Holowinsky will work only for surfaces with a high degree of symmetry and with constant negative curvature. (The full conjecture would apply to surfaces with curvature that varies from point to point.) Even so, according to Rudnick, the two mathematicians have pulled off a rare feat. "It's one of those rare cases where you can get around the Riemann Hypothesis," Rudnick says. "In many problems where the Riemann Hypothesis can be applied, we're so far away from getting around it. But here was one case where we needed only a little bit extra, and Soundararajan found a short cut."

**The whole story is a beautiful example of the interplay between physics and mathematics. The physical problem motivated Sarnak and Rudnick to propose a number-theoretic analogue that might never have occurred to them without the stimulus from quantum chaos.**

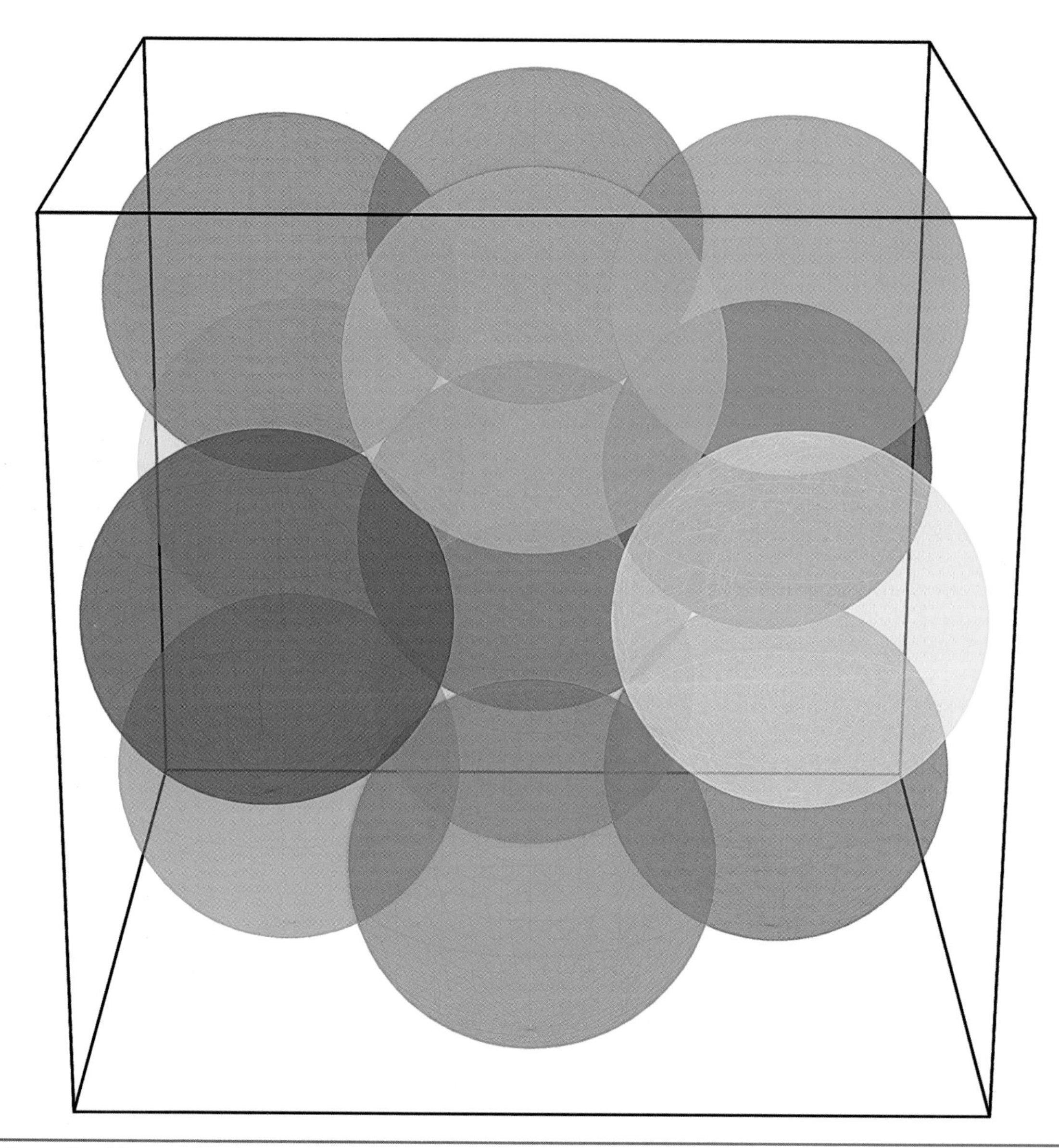

**Sphere Packing.** *The densest packing of equal-sized spheres, called the face-centered cubic packing. It is highly symmetric (a lattice) and also remarkably porous. Two questions still motivate much research in packings: Under what circumstances is the densest packing a lattice? Are spheres the least efficient bodies for packing space? (Figure courtesy of elizabeth r. chen. ©2010 elizabeth r. chen.)*

# 3-D Surprises

IT IS NO SECRET THAT CONTEMPORARY MATHEMATICAL research is often difficult to grasp. Some problems involve exotic kinds of algebra. Others involve spaces of more than three dimensions—sometimes *way* more than three. And even when mathematicians work in three-dimensional space, they are often studying phenomena like fluids or vector fields—things that are impossible to hold in your hand.

Rarely does a sudden and unexpected breakthrough occur in our understanding of ordinary, solid, three-dimensional objects. But it has happened twice recently. In 2008 and 2009, scientists unlocked a variety of new ways to pack tetrahedra (i.e., pyramids with a triangular base) extremely densely. In a flurry of new papers, sometimes written only days apart, four different teams of researchers smashed through an intellectual logjam on a century-old problem.

A couple of years earlier, two engineers in Hungary discovered a new three-dimensional object with a unique talent. It is the first homogeneous, self-righting (and self-wronging!) object. That is, it has only one stable balance point and one unstable balance point. Coincidentally, their creation, which they dubbed the Gömböc, vaguely resembles a tetrahedron, albeit with curvy sides (see Box, **"The Incredible Self-Righting (and Self-Wronging) Object,"** page 102).

So who needs 126 dimensions, when even our familiar three-dimensional space harbors such surprises?

**Rarely does a sudden and unexpected breakthrough occur in our understanding of ordinary, solid, three-dimensional objects. But it has happened twice recently. In 2008 and 2009, scientists unlocked a variety of new ways to pack tetrahedra (i.e., pyramids with a triangular base) extremely densely.**

## Efficiency and Symmetry

The problem of packing tetrahedra begins with a misunderstanding. In the fourth century B.C., Aristotle wrote, "It is agreed that there are only three plane figures which can fill a space, the triangle, the square, and the hexagon, and only two solids, the pyramid and the cube." It is quite clear that the pyramid he means is the regular tetrahedron, because in this passage he was trying to refute the attempts of other Greek philosophers to find a correspondence between the elements (such as air and fire) and the Platonic solids (tetrahedron, cube, octahedron, dodecahedron, and the 20-sided icosahedron).

Aristotle was correct in arguing that elements have nothing to do with Platonic solids, but he was wrong about tetrahedra. In fact, you cannot fill all of space with these four-sided figures. If you try to fit them together, you will find that you can place five in contact around an edge, but there is a gap of 7.36 degrees left over. Or you can fit twenty around a vertex, but again the fit is loose. (See Figure 1)

Once you accept that a tetrahedron packing cannot be 100 percent efficient, it is natural to ask how efficient it can be. This exact question was raised by David Hilbert in the most famous mathematical problem list of all time. In 1900, at the second International Congress of Mathematicians in Paris, Hilbert presented an ambitious list of 23 questions that he considered the most important unsolved problems in mathematics. Hilbert's eighteenth problem reads, in part: "How can one arrange most densely in space an infinite number of equal solids of given form, e.g., spheres with given radii or regular tetrahedra with given edges (or in prescribed position), that is, how can one fit them together that the ratio of the filled to the unfilled space may be as great as possible?"

This problem can be read as a manifesto for research on packing problems in general. But it also contains two very specific challenges: To find the best possible packing of spheres and the best possible packing of tetrahedra. (Continued on p. 105)

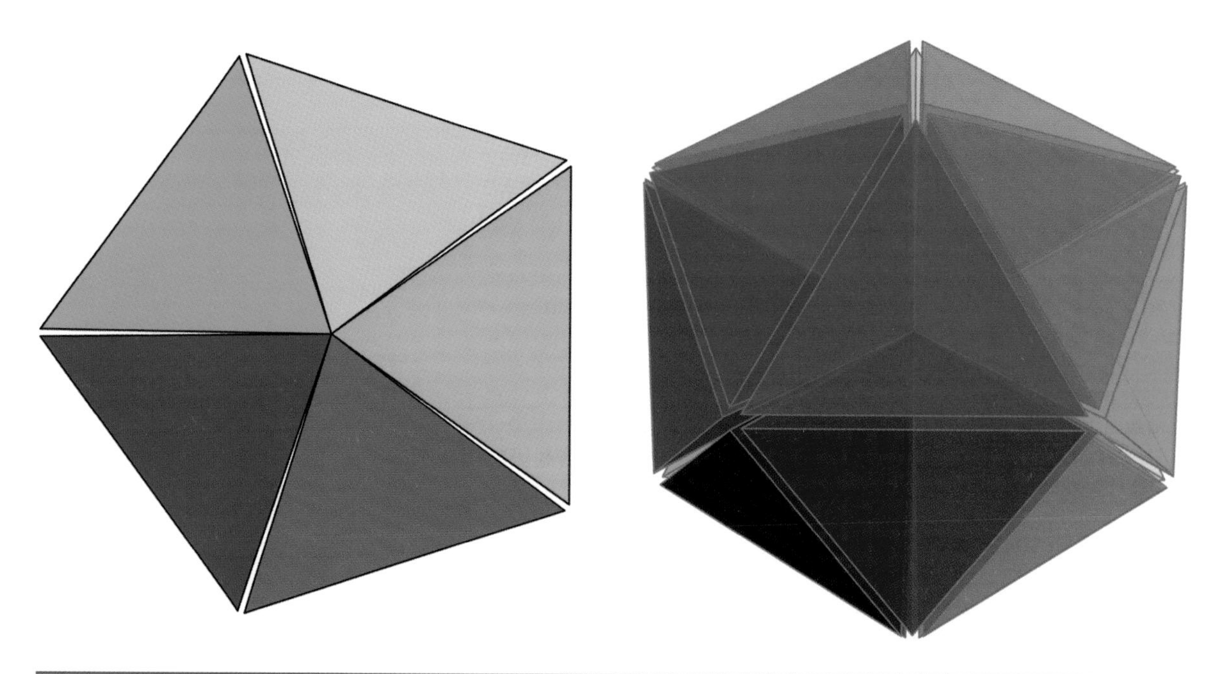

**Figure 1.** *Regular tetrahedra cannot fit snugly around an edge (left) or a vertex (right). (Figures courtesy of elizabeth r. chen. ©2010 elizabeth r. chen.)*

**The Incredible Self-Righting (and Self-Wronging) Object**

In 1974, a new toy from Hungary took the world by storm—Rubik's cube, a twistable and turnable object that seemed to defy the laws of physics. More than thirty years later, another mathematical toy emerged from the fertile mind of Hungarian engineers Gábor Domokos and Péter Várkonyi—the Gömböc, a three-dimensional object that has only one point of stable equilibrium and one point of unstable equilibrium.

Though nowhere nearly as popular as Rubik's cube, the Gömböc exerts the same sort of "How is that possible?" fascination. It's a roughly tetrahedral shape, with a high arched back like that of a stegosaurus or, as Domokos points out, certain tortoises. If you put one down sideways, you will

**Gömböc 1.** *A family of Gömböcs, manufactured by a computerized numerical control (CNC) process from an aluminum-magnesium-silicon alloy. (Photo by István Oravecz and courtesy of Gábor Domokos.)*

see it wobble around and sometimes come very close to settling in a non-stable position, before it suddenly "changes its mind" and heads for its unique stable equlibrium.

The Gömböc also has only one unstable equilibrium point (on top of the arched back). So it is not only self-righting but also "self-wronging." For example, if you made one that was lighter than air, but tethered to the ground, it would always rotate around to the same "wrong" orientation. Because any 3-dimensional convex body must have at least one stable and one unstable equilibrium (the points that are at maximum and minimum distance from the center of mass) the Gömböc has the fewest possible balance points.

Self-righting toys have been constructed before, but they have always been aided by a heavy weight at one end. The Gomboc is the first object of its kind that is homogeneous; for example, you can make it out of glass to see that there are no hidden counterweights.

As Domokos says, the idea for the Gömböc came when he attended a lecture in 1995 by the famous Russian mathematician Vladimir Arnold, about the so-called "four vertex theorem" and its relatives. Domokos (together with Andy Ruina and Jim Papadopoulos) had proved a similar statement of this type: that a convex planar shape must have at least two stable equilibria and two unstable ones. In fact, this is not too hard to see. If you draw the shape as the graph of a function $r = f(\theta)$ in polar coordinates, then the function $f$ must have at least one minimum and at least one maximum. (This is an elementary theorem of topology.) If you assume that it has *only* one minimum, say at $\theta = 0$, then you can find an interval $[-\alpha, \pi - \alpha]$ of length $\pi$, which contains 0, and has the property that $f(\theta)$ is smaller at any angle $\theta$ inside the interval than outside. It then follows

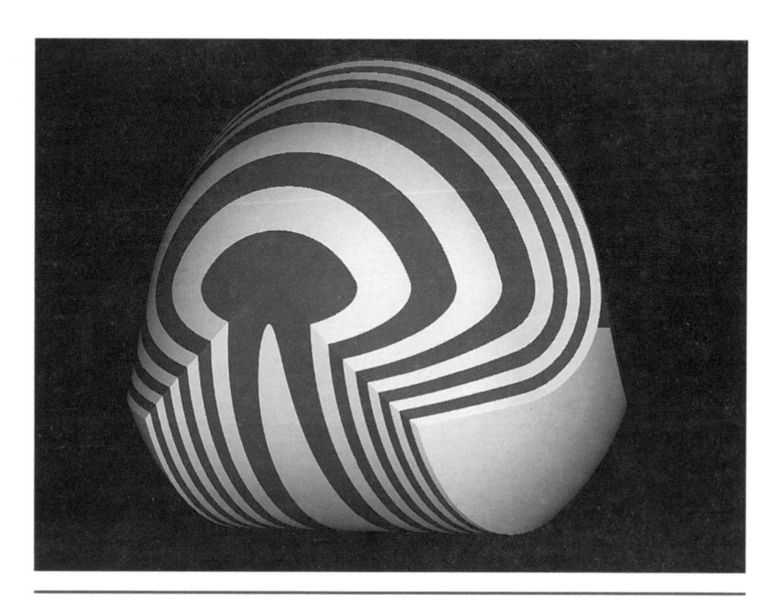

**Gömböc 2.** *A computer-drawn model of the Gömböc reveals its tennis ball-shaped structure. Shaded bands represent points that are equidistant from the Gömböc's center of mass. The unique stable balance point lies at the center of the innermost white stripe (the closest point to the center of mass). The unique unstable balance point lies at the top of the Gömböc, at the point which is farthest from the center of mass. When placed on its side, the Gömböc nearly balances at the center of the dark region at the front, but eventually tips over and comes to rest in the orientation shown. (Figure courtesy of Péter Várkonyi.)*

that the object is "lopsided". If you draw a line through the center of mass with two rays pointing in the directions $\theta = -\alpha$ and $\theta = \pi - \alpha$, then it separates the object into two halves, one of which is fatter than the other. This is inconsistent with the assumption that the line passed through the center of mass. By a *reductio ad absurdum*, we can conclude that the assumption of only one minimum was impossible.

If you try to apply the same argument to a three-dimensional body, it doesn't quite work. You can still separate the body into a "fatter" half and a "thinner" half. If the two halves were hemispheres, then the argument above would be fine. But if the two halves can embrace each other like the two halves of a tennis ball, it is not clear whether the "fatter" half will force the three-dimensional body out of balance.

Despite the unproven, but common, belief that a three-dimensional body must also have at least four equilibria, Arnold speculated that a body with only two does exist. Domokos tried to find such an object, but he was defeated by its elusive properties. It could not be too flat, because (like an M&M candy) it would have two stable balance points at the top and bottom. It could not be too elongated or thin, because it would have two unstable balance points, like a cigar or an egg (which balances unstably on either end). In fact, it seemed very much like a sphere—but a sphere wouldn't do, because it balances *everywhere*. The self-righting property is very sensitive to small changes too; for example, you can make an egg balance on one end by crushing the end slightly. Domokos even resorted to collecting 2000 pebbles on a beach in hope of finding one that would work. Alas, none of them did.

But finally, around 2005, Domokos got two of his Ph.D. students interested in the problem. One tried to prove the body did exist, and one tried to prove it didn't. "I was the luckier one," Várkonyi says modestly. He and Domokos found a tennis ball shape that worked, with one of the interlacing halves bulging out more than the other. Not surprisingly, the shape was very close to spherical. Várkonyi estimates that the bulging side is only about 0.1 percent farther from the center of mass than the other one. To the naked eye, such an object would look completely like a sphere.

However, once they knew that such a shape was possible, and that the "tennis ball" idea worked, they wanted to design a figure whose irregularity was more visible. Finally they found a better solution, which was patched together with pieces of cylinders, ellipsoids, and cones, and drawn with commercial 3-D rendering programs. They could then use a prototyping machine to make a real, physical example. The resulting body, which they named the Gömböc (a creature from a Hungarian folk tale), has been confirmed numerically to have the self-righting, self-wronging property. Instead of trying to write out a formal mathematical argument, as they had done for the original (nearly spherical) self-righting shape, Várkonyi and Domokos decided this time to let nature provide the proof. Where the Gömböc is concerned, seeing is believing!

(Continued from p. 102) It is very odd, given the great prestige of Hilbert's problems, that mathematicians for the most part paid attention only to the first part of this problem. An obvious candidate existed for the densest packing of spheres: The face-centered cubic packing, an arrangement that is common in crystals and is also familiar to anyone who has seen a stack of oranges in the grocery store (see Figure **"Sphere Packing,"** page 100). Way back in 1611, Johannes Kepler had conjectured that this was the most efficient way to pack spheres. Curiously, Hilbert did not mention Kepler's conjecture, perhaps because he was not sufficiently convinced that it would turn out to be true. Thomas Hales finally confirmed it in 1998 (see *What's Happening in the Mathematical Sciences*, volume 5), with a highly computer-intensive proof. As a result, we now know that there is no way to fill more than 74.05 percent of space with equal-sized spheres.

Kepler's Conjecture, now Hales' Theorem, is a seductively beautiful type of theorem. It says that *optimality* leads to *symmetry*. The whole history of packing problems is a battle between this "Panglossian" viewpoint on the world and a more "Discordian" viewpoint that prefers disorder to order. In the case of sphere packing, harmony wins. The spheres form a lattice, as orderly as the atoms in a crystal.

However, the situation for tetrahedra is much more murky. Finding the densest *lattice* packing is comparatively easy, because the lattice can be completely described by three basis vectors $\mathbf{v}_1$, $\mathbf{v}_2$, and $\mathbf{v}_3$. Every single tetrahedron in the lattice can be obtained from the first one by translating it an integer number of steps in each of these three directions. Thus the density of the packing is "merely" a function of nine variables (the three coordinates of each of the three basis vectors). By contrast, an arbitrary packing depends on infinitely many variables (the location and orientation of each unit in the packing). Consequently, it is vastly easier to find the densest *lattice* packing than to find the densest packing *unconditionally*.

In 2006, Salvatore Torquato, a chemist at Princeton University, teamed up with John Conway, a well-known Princeton mathematician, to find *non-lattice* packings of tetrahedra that were nearly twice as dense as the best lattice packing (see Figure 2). This result showed that there was a vast difference between spheres, whose densest packing is a lattice, and tetrahedra, whose densest packing is far from being a lattice.

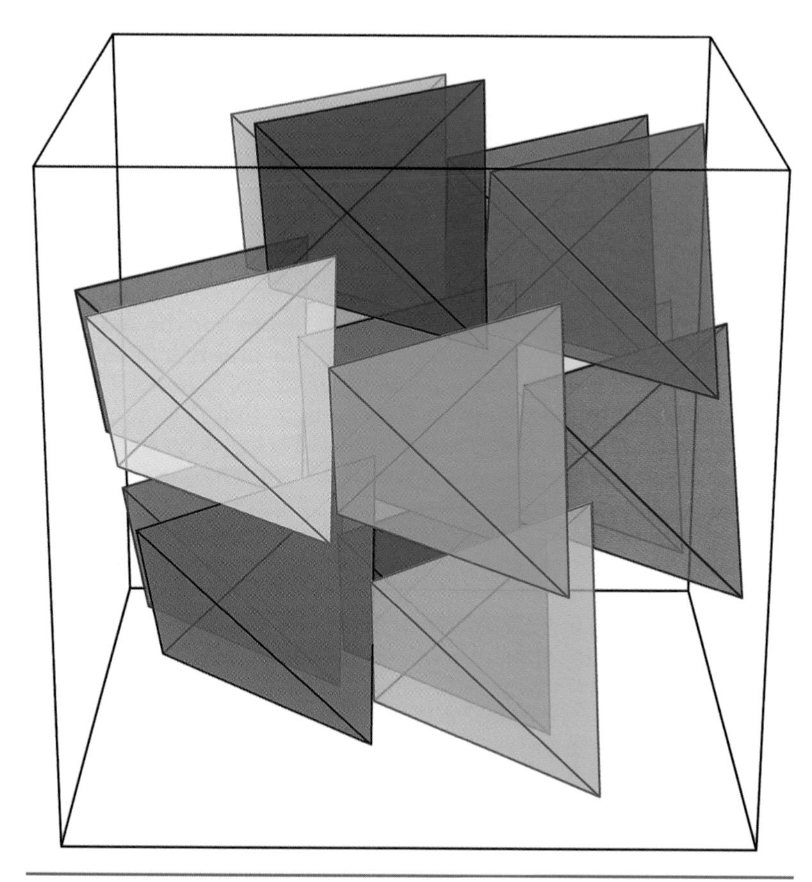

**Figure 2.** *The densest lattice packing of tetrahedra. This packing is far less dense than the non-lattice packings shown below. (Figure courtesy of elizabeth r. chen. ©2010 elizabeth r. chen.)*

The idea behind their construction was to place 20 tetrahedra around a vertex (as in Figure 1 (right)), jiggle them so that they fit in the smallest possible regular icosahedron, look for the densest lattice packing of these icosahedra, and then possibly jiggle some more to take advantage of the remaining gaps between tetrahedra. In this way they arrived at a packing with a density of about 71.75 percent. It could not be a lattice packing because all of the tetrahedra in a lattice have the same orientation, while the tetrahedra in Torquato and Conway's construction had 20 different orientations.

At this point, it seemed likely that the density could be improved but not by much. Thus, Conway and Torquato conjectured that the regular tetrahedron might actually be the least efficient packer of space, on the grounds that the tightest packing of tetrahedra seemed to be less dense than that of spheres (74.05 percent). Curiously, this directly contradicted an earlier supposition by Stanislaw Ulam, who believed that the sphere was the least efficient packer. The truth was that nobody knew.

**Sal Torquato.** *(Photo courtesy of Sal Torquato.)*

## The Pace Quickens

Meanwhile, unbeknownst to Torquato, a graduate student at the University of Michigan had already been working on the tetrahedron problem for a year, as her dissertation project. Beth Chen says that she chose the problem from a list of topics suggested by her advisor, Jeff Lagarias, because "I would not be happy unless I could draw beautiful pictures." Her Ph.D. thesis (completed in 2010) is called "A Picturebook of Tetrahedral Packings," and the title is a very accurate description of the contents: 200 pages with thousands of pictures and very little conventional text.

Chen's overall strategy was quite similar to Torquato and Conway's: First, look for *locally dense* packings of tetrahedra, called "clusters"; second, construct optimal crystallographic packings of those clusters. On the spectrum between harmony and discord, this can be described as mild discord.

However, Chen did one thing that Torquato and Conway didn't: She experimented systematically with a variety of different clusters, not just the 20-tetrahedron cluster shown in Figure 1 (right). For example, the 5-tetrahedron cluster of Figure 1 (left), which Torquato calls a "wagon wheel," would be a good starting point, because it has a higher local density than the 20-tetrahedron cluster (97.96 percent compared to 87.74 percent). However, pentagons don't fit together very well.

Instead, Chen found a 9-tetrahedron cluster (see Figure 3, next page) that fits together very nicely. Here is how you construct the cluster: You start with one tetrahedron, which we can call the central tetrahedron. Then you attach four tetrahedra to it, which form a sort of claw that latches onto two faces of the central tetrahedron. Next you attach another group of four tetrahedra that latch on to the two opposite faces. Note that the second claw makes a 90-degree angle to the first claw. Another way to visualize this object is to think of it as two wagon wheels with exactly one tetrahedron of overlap.

Having constructed the individual 9-tetrahedron clusters, it remains to explain how to pack them together. Chen arranged them into what Greg and Wlodzimierz Kuperberg call a "double

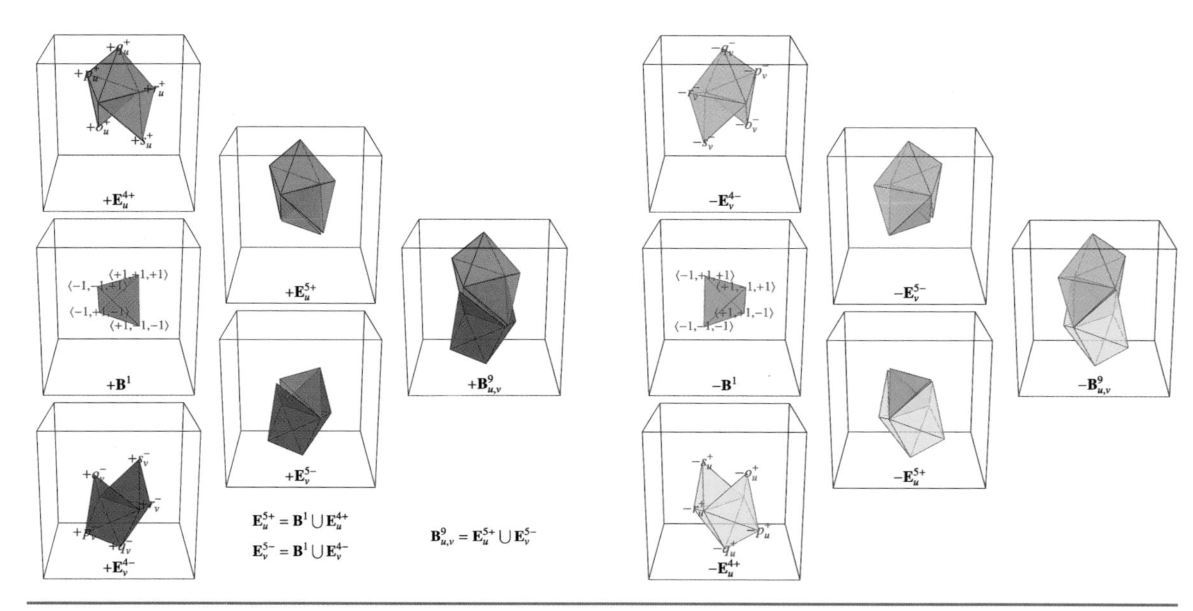


**Figure 3.** *(left) The basic nine-tetrahedron unit for Chen's first packing of tetrahedra. Each unit is composed of two "wagon wheels" (red and blue) twisted by 90 degrees relative to one another, which share one tetrahedron (pink). (right) A central inversion of the basic unit, with aqua, green, and yellow corresponding to blue, pink and red respectively. (Figure courtesy of elizabeth r. chen. With kind permission from Springer Science+Business Media: "A Dense Packing of Regular Tetrahedra," Discrete & Computational Geometry,* **40**, *no. 2 (2008).)*

lattice." (This term was not used in Chen's thesis.) A double lattice can be created from a single lattice by an operation called central inversion, which maps every point in space $(x, y, z)$ to its opposite point $(-x, -y, -z)$. Central inversion turns out to be critically important in packing polyhedra, because it allows for face-to-face contact. The problem can be seen most visibly in the lattice packing of Figure 2. As long as we keep the tetrahedra in their original, "positive" orientations, we can never form a face-to-face contact of two tetrahedra—only edge-to-edge, edge-to-face, and vertex-to-face contacts, which waste a great deal of space. The 20-unit clusters of Torquato and Conway, or the 9-unit clusters of Chen, without central inversion, allow for relatively few face-to-face contacts. By contrast, a positively-oriented cluster and its negatively-oriented central inversion can be translated around to have many different face-to-face contacts. That is exactly the fact that Chen's packing exploited. As seen in Figure 4, a layer of negatively-oriented clusters (made up of blue, pink, and red tetrahedra) fits very nicely between two layers of positively-oriented clusters (made up of yellow, green, and light blue tetrahedra). As it turns out, the 90-degree twist between the upper claw and lower claw in each cluster also facilitates the snug fit. (See Figure 4.)

Having described the double-lattice arrangement of the clusters, the only remaining thing that Chen had to do was find the best way to wiggle the "claw" units to make the double lattice as dense as possible. (Remember that the tetrahedra in the wagon wheels do not fit snugly; there is slightly more than 7 degrees of "play" that can be exploited to make the packing denser.) This

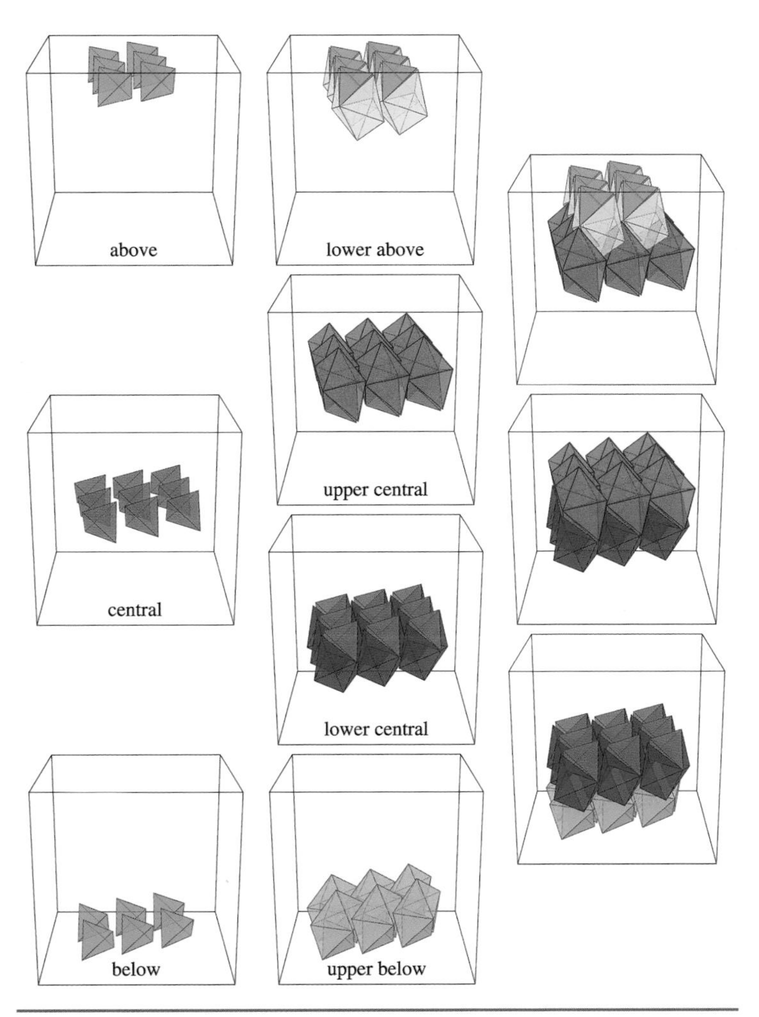


**Figure 4.** *The assembly of Chen's nine-tetrahedron units into a double lattice packing. Every unit can be mapped to any other unit by either a translation or a central inversion that preserves the entire arrangement. However, the packing is not a double lattice on the level of tetrahedra. (Figure courtesy of elizabeth r. chen. With kind permission from Springer Science+Business Media: "A Dense Packing of Regular Tetrahedra," Discrete & Computational Geometry,* **40**, *no. 2 (2008).)*

part required computer calculations. With the claws in their default positions the packing density was 77.72 percent, and wiggling them increased the density to 77.86 percent.

Chen's paper, which appeared in *Discrete and Computational Geometry* in 2008, smashed the "sphere-packing" barrier; it was now clear that tetrahedra could be packed a good deal more efficiently than spheres. Meanwhile, Torquato had not been standing still. He and graduate student Yang Jiao had developed a computer program, called the "adaptive shrinking cell" method, which allowed them to take clusters of Chen's clusters and jiggle all of the tetrahedra in these super-clusters by a little bit. The result was a less ordered packing with a slightly higher packing density of 78.20 percent. Dis-

**elizabeth r. chen.** *(This photograph is used with permission from the Ann Arbor Chronicle.)*

cord was starting to win out over harmony. "It was clear that tetrahedra were very weird objects, and that we didn't really have any handle on what was going on," Torquato says.

### Quasicrystals and Sudoku Puzzles

The next two groups to enter the tetrahedron-packing sweepstakes arrived at the problem in very different ways.

At the University of Michigan, chemical engineering professor Sharon Glotzer was working in 2008 on a project for the U.S. Air Force on materials assembled from tetrahedral nanoparticles. As a simple model, she and physicist Peter Palffy-Muhoray of Kent State University thought it would be informative to look at the case where the particles were infinitely hard. At this point she did not know about Torquato's work and did not know that Chen had been working on tetrahedron packings at the very same university.

Unlike Chen and Torquato, Glotzer (along with postdoc Michael Engel and students Amir Haji-Akbari and Aaron Keys) took a thermodynamic approach to the problem. They simulated a random fluid made out of tetrahedra, using a version of the Metropolis algorithm (see "Instant Randomness," page 56) tweaked in such a way that the probability of acceptance of a random move would remain roughly constant, around 30 percent. After giving each of the $n$ particles in their simulation a chance to move randomly, they would then try to shrink the size of the box they were contained in, altering the shape of the box too (if necessary). The last step is a standard add-on to the Metropolis algorithm, called isothermal-isobaric Monte Carlo, and is similar to Torquato's "adaptive shrinking cells."

In essence, the idea was to put the fluid under higher and higher pressure, and let it find its own way to a dense packing. The original problem, to find the maximum density possible, is the infinite-pressure limit of this process.

"What we found was that a stunningly unexpected structure formed spontaneously—a quasicrystal," Glotzer says. Quasicrystals are a state of matter that was discovered in the 1980s, which is different from a crystal because it lacks a periodic structure, but is nevertheless highly patterned. One of the diagnostic features of a quasicrystal is that it diffracts light into a sharply focused pattern of dots with a "forbidden" symmetry that is not allowed in a lattice. The quasicrystals discovered in the 1980s had 5-fold symmetry. The simulated quasicrystal in the computers of Glotzer's students had 12-fold symmetry. Now the tug-of-war between discord and harmony had veered off into a new and completely unexpected direction!

Curiously, the thermodynamically random tetrahedron packing doesn't even have to be very dense to start forming a quasicrystal; it begins happening around a density of 50 percent. Glotzer and her colleagues were able to compress the quasicrystals to a density as high as 83 percent, moving well past both Torquato and Chen. But then they discovered they could do even better by taking a "quasicrystal approximant." Figure 5 shows the densest packing they could find, a crystalline structure formed from an 82-cell cluster. When

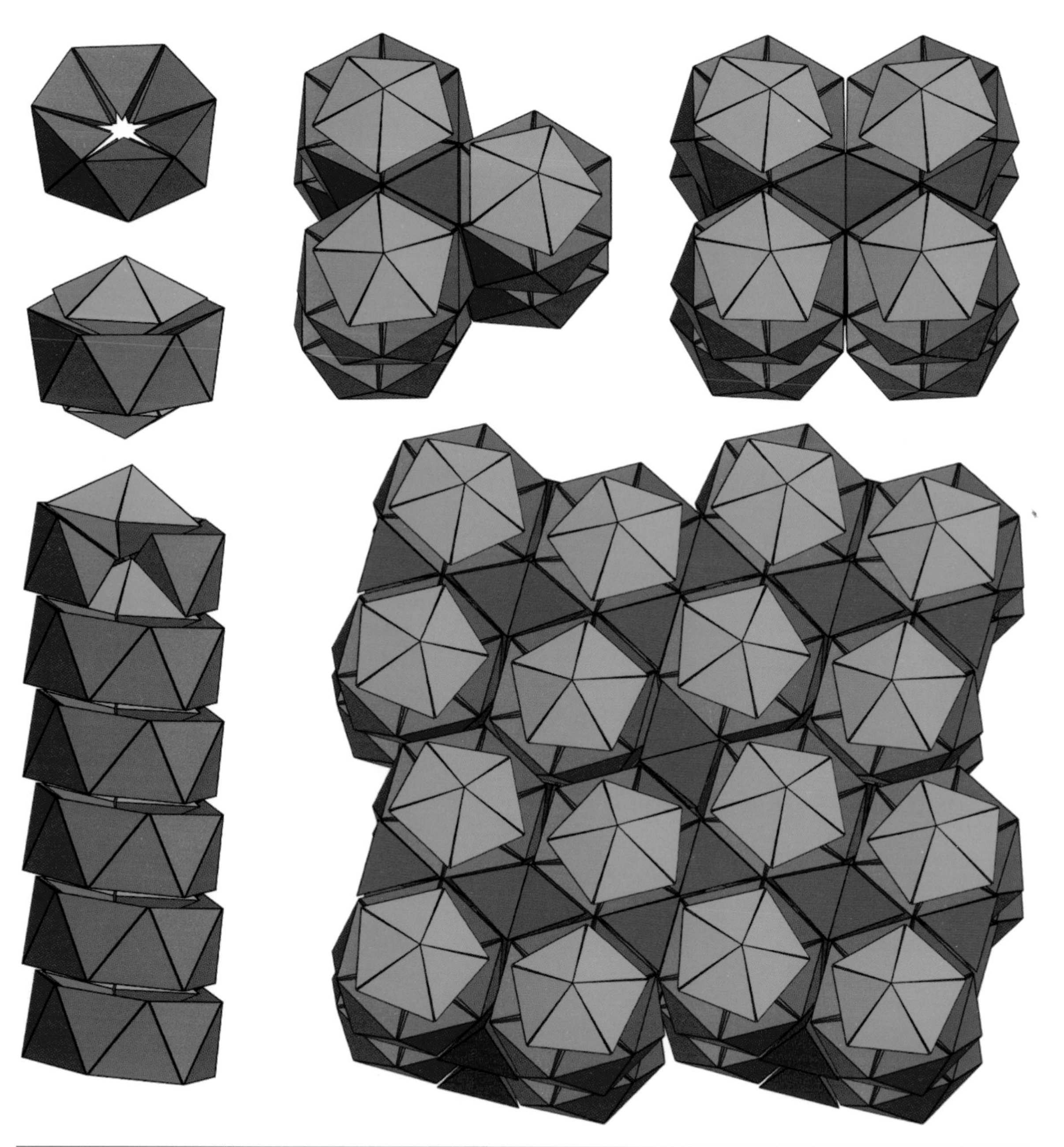

**Figure 5.** *A packing discovered by Glotzer et. al., based on an 82-tetrahedron unit, which fills 85.03 percent of space. The top and sides of this figure illustrate different motifs that occur repeatedly in the complete packing. (Figure courtesy of Michael Engel. Reprinted by permission from Macmillan Publishers Ltd: "Disordered, quasicrystalline and crystalline phases of densely packed tetrahedra," Nature, A. Haji-Akbari, M. Engel, A. S. Keys, X. Zheng, R. G. Petschek, P. Palffy-Muhoray, S. C. Glotzer, ©2009.)*

the clusters are pieced together, they form the same patterns Glotzer's group had seen in the quasicrystals—five-unit wagon wheels, nestled inside "logs" that are formed by double rings of six tetrahedra, and interspersed with twisting filaments called tetrahelices. The packing shown in Figure 5 fills 85.03 percent of space. This structure was *not* formed thermodynamically—it is a creation of the human mind, inspired by thermodynamics.

But Glotzer's group would not have the densest packing for long. In November, a team from Cornell University reported a packing whose density was exactly 100/117, or 85.47 percent. (See Figure 6a and 6b.) This result came completely out of left field. First, it reversed the trend toward ever more complicated packings with less and less symmetry. The Cornell packing used a basic cluster of only 4 tetrahedra, with no computer-aided jiggery and pokery. Not only that, it was extremely symmetric. In fact, it is a double lattice in the sense of Kuperberg and Kuperberg: a union of a lattice with its central inversion. It was, in short, a beautiful arrangement that everybody, from Conway to Torquato to Chen to Glotzer, had overlooked.

"It was quite shocking," Torquato says. "It was readily available to us, but we had just never considered such small packings. Everything had suggested that the denser the packing, the more particles you needed. That was blown out of the water by this Cornell paper!"

The second remarkable thing was the way that the Cornell team—grad students Yoav Kallus and Simon Gravel, and their advisor Veit Elser of the physics department—discovered the packing. It grew out of a method that Elser has pioneered for solving constrained optimization problems, called the "difference method."

Elser has also used the difference method to solve Sudoku puzzles, and this actually provides a nice illustration of how the method works. A Sudoku puzzle has two constraints, which we can call A and B:

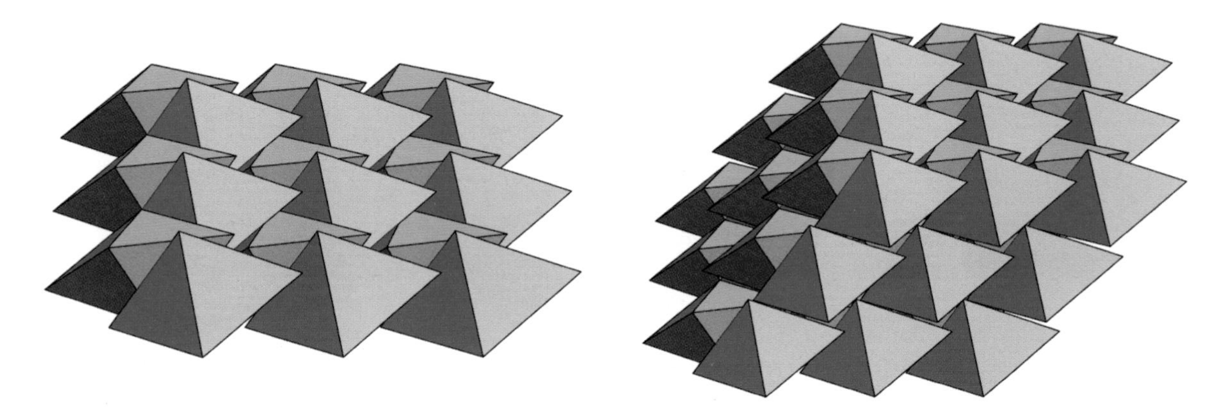

**Figure 6a.** *A packing of tetrahedra, discovered by Kallus et. al., that fills 85.47 percent of space. It is a lattice of 4-tetrahedron units and a double lattice of dipyramids. It also has an extra rotational symmetry, so that every tetrahedron is equivalent to every other one by either a translation, a central inversion, or rotation. (Figure courtesy of Yoav Kallus. With kind permission from Springer Science+Business Media: "Dense Periodic Packings of Tetrahedra with Small Repeating Units," Discrete & Computational Geometry,* **44***, no. 2 (2010). ©2010.)*

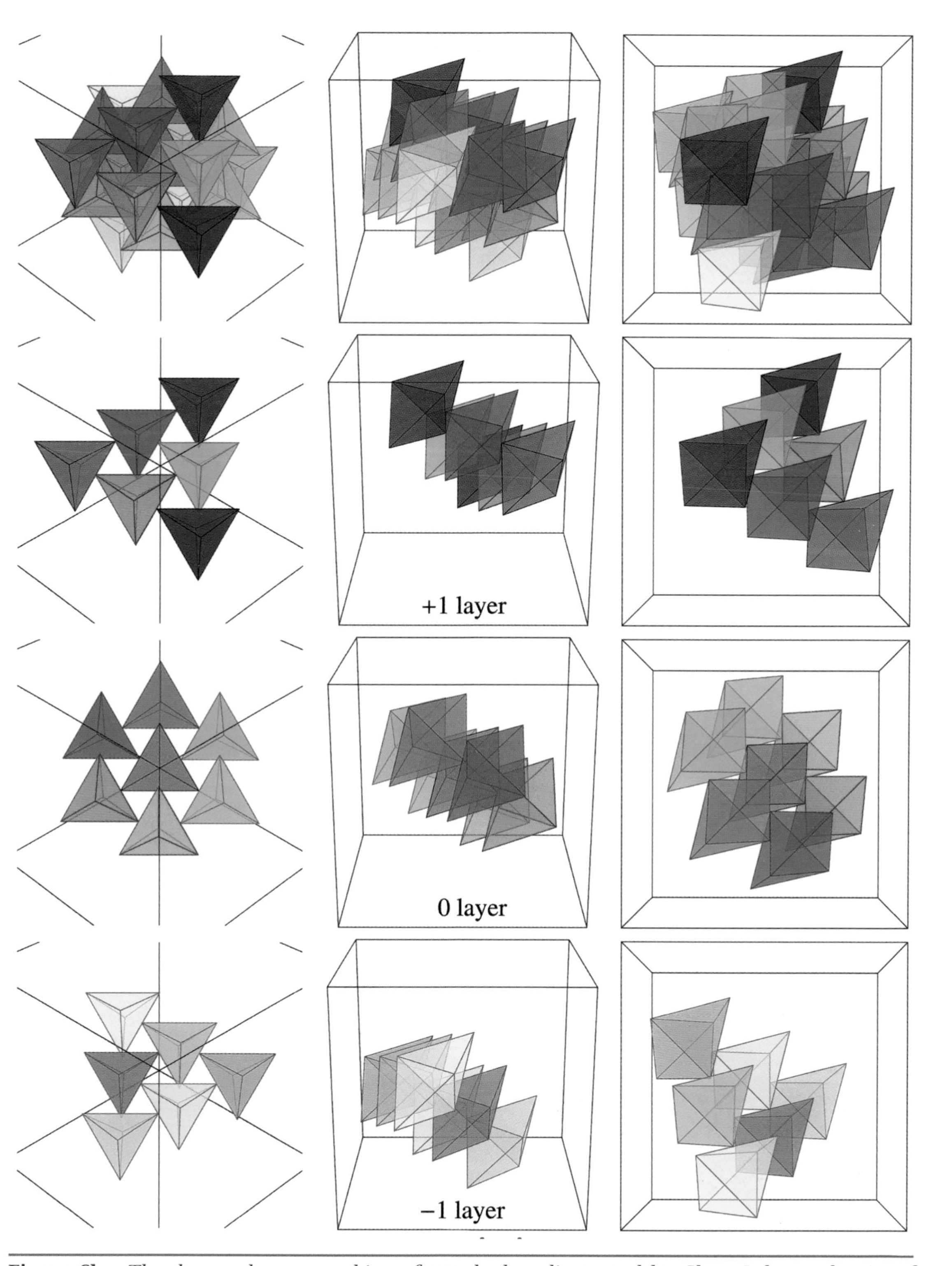


**Figure 6b.** *The densest known packing of tetrahedra, discovered by Chen. It has a density of 4000/4671; that is, it fills 85.63 percent of space. It is virtually indistinguishable from Kallus's packing to the naked eye, but it is not rotationally symmetric. (Figure courtesy of elizabeth r. chen. ©2010 elizabeth r. chen.)*

**Yoav Kallus.** *(Photo courtesy of Yoav Kallus.)*

Constraint A: Each number from 1 to 9 appears once in each row and column.

Constraint B: Each number from 1 to 9 appears once in each $3 \times 3$ block.

Let us say that a "generalized solution" of a Sudoku puzzle is a $9 \times 9 \times 9$ cube of numbers $x_{ijk}$, which are not necessarily integers. For example, if $x_{276} = \frac{1}{2}$, then you write $\frac{1}{2}$ of a number "6" in the second row and seventh column of the puzzle. Or if $x_{418} = -1/3$, you write $-1/3$ of a number "8" in the fourth row and first column. In other words, a generalized solution allows for a superposition of fractional states (an idea that may seem outlandish, but is quite normal in quantum mechanics). Of course, for a classical solution no fractions and no superpositions are allowed: You must have a single 1 in each row and column of the $x_{ijk}$ cube and 0's elsewhere.

Starting with any generalized solution $x_{ijk}$, it is easy to compute its "projection" to the space of classical solutions that satisfy Constraint $A$, denoted $P_A(x_{ijk})$. Likewise, it is easy to compute its projection to the nearest classical solution that satisfies constraint $B$, denoted $P_B(x_{ijk})$. A naïve way to solve the puzzle, then, is to alternately compute the projections $P_A$ and $P_B$ until you converge to a solution:

$$P_B(P_A(P_B(P_A(\ldots(x_{ijk})\ldots)))).$$

Unfortunately, the naïve method doesn't work. You can get stuck oscillating back and forth between two points, one satisfying constraint A and one satisfying constraint B but neither satisfying both of them. However, a slightly cleverer approach, proposed by James Fienup of the University of Rochester, uses instead the iterations of a single map:

$$D(x) = x + P_A(2P_B(x) - x) - P_B(x).$$

If $x$ is a fixed point of $D$ (i.e., $D(x) = x$), then it is easy to see that

$$P_A(2P_B(x) - x) = P_B(x).$$

In this case, $P_B(x)$ lies in the image of both projection maps $P_A$ and $P_B$, and therefore it satisfies both constraints, $A$ and $B$.

Unfortunately, Fienup's map $D$ still doesn't always work because $D(D(D(\ldots(x)\ldots)))$ does not necessarily converge to a fixed point. But Elser has devised a modified version of $D$, called the "difference map," with much better convergence properties. For example, it can solve any solvable Sudoku puzzle, without fail, in a finite number of steps.

When they saw Torquato's *Nature* paper, Kallus says, "We thought, here is a cool problem that people are interested in, let's try to solve it with our method." Their approach, which Kallus named the "divide and concur" algorithm, was to treat each tetrahedron as a superposition of many virtual tetrahedra. (That is the "divide" step.) Constraint $A$ says that certain tetrahedra don't overlap. Constraint $B$ says that all of the virtual tetrahedra that do overlap are in fact identical. (That is the "concur" step.)

By using a computer to iterate the corresponding difference map $D$, Kallus arrived at a packing that strongly resembled Figure 6. It is a lattice packing with a generating cluster of only 4 tetrahedra—a "bipyramid" (two tetrahedra stuck together face-to-face) and its central inversion. It was just a simpler version of Chen's earlier line of attack, which consisted of a 9-unit cluster and its central inversion. This one uses a 2-unit cluster and its central inversion. The difference is that the bipyramid is an even denser local configuration than a 9-unit cluster: its density is 100 percent!

"I was extremely surprised," says Kallus. "I didn't expect such a simple solution, because the trend in these dense packings up to then was that each denser packing was more complicated than the previous one." The computer's solution was so simple that they were able to throw away the "divide and concur" algorithm and calculate explicitly, using linear algebra, where each tetrahedron should be.

But here they made a fatal, Panglossian assumption. They assumed the solution was more symmetric than it actually was. As mentioned before, they correctly surmised that the packing was a double lattice. (By this point, Kallus had discovered the paper by the Kuperbergs that introduced this term.) Thus, by definition, it had infinitely many translational symmetries and one central inversion symmetry. But their computer drawings also appeared to have a rotational symmetry. Under that assumption, they worked out the density and found it was 100/117. Their calculation was correct: a packing with the stated symmetries and density does exist. But it is not the densest.

Torquato was the first to realize that the extra rotational symmetry was not necessary. As noted before, central inversion symmetry is very useful for a packing to have, because it allows face-to-face contact. But rotational symmetry does not serve any obvious purpose. By dispensing with this assumption, Torquato found a two-parameter family of packings, whose maximum density was ever so slightly greater than the Cornell packing, at 85.55 percent. The two packings are so similar that they are almost impossible to distinguish by the naked eye.

**As noted before, central inversion symmetry is very useful for a packing to have, because it allows face-to-face contact. But rotational symmetry does not serve any obvious purpose.**

Only days later, Chen realized that Torquato had missed a parameter: the Cornell packing is actually part of a three-parameter family of packings, with the three parameters corresponding to subtly different positions of the two bipyramids within a cluster. Maximizing the density in this family (a simple, three-variable calculus problem!), she found a configuration with a density of 4000/4671, or 85.63 percent—just a smidgen over Kallus' 85.47 and Torquato's 85.55 percent, but enough to take the record away.

Will this record hold up? It's hard to say. Chen has proved that this is the densest lattice pattern with a 4-unit cluster and with central inversion symmetry. Engel, and Glotzer have tried "jiggling" the solution on the computer, but unlike many of the previously proposed solutions, jiggling doesn't make it any denser. They have also done an empirical search for all of the densest clusters with 1 to 16 units, using the same thermodynamics-based simulation algorithm that they had used previously. Again, nothing better has turned up. In fact, the densest clusters with 8, 12, or 16 units are simply repetitions of Chen's 4-unit packing.

But, on the other hand, the history of the problem suggests that further surprises would not be much of a surprise. Perhaps a perfect quasicrystalline packing, or a higher-order quasicrystalline approximant, may yet surpass Chen's packing, and the balance in the packing problem will swing once again toward disorder rather than harmony.

**Michael Engel and Sharon Glotzer.** *(Photo courtesy of Michael Engel.)*

Meanwhile, the field of packings in general seems to be just getting started. Random packings—not only of tetrahedra and spheres but other shapes—are far from understood. The phase changes that appear in the thermodynamic approach, with "fluid" disordered structures at low pressures, quasicrystals at higher pressures, and periodic structures at infinite pressures, beg to be explored. Are there any shapes for which the densest packing is actually disordered? In this regard, Torquato and Jiao have conjectured that for centrally symmetric polyhedra, the densest packing will always be a lattice (see Figure 7). Henry Cohn of Microsoft Research noted in a comment to their *Nature* article that this could be a "Kepler's conjecture for the twenty-first century."

Whatever the final outcome, the progress on tetrahedron packing in less than four years has been nothing short of amazing. As Kallus wrote in his paper, "The regular tetrahedron is no longer [an] outcast." It has gone from being, seemingly, one of the clumsiest packers of space to being very efficient. The variety of approaches used for the problem, from pure linear algebra to thermodynamics to a souped-up Sudoku-solving algorithm, has also been remarkable. The computer has played a major role, and yet in every case the human mind has been able to extract patterns that the computer did not detect. And the story isn't over yet!

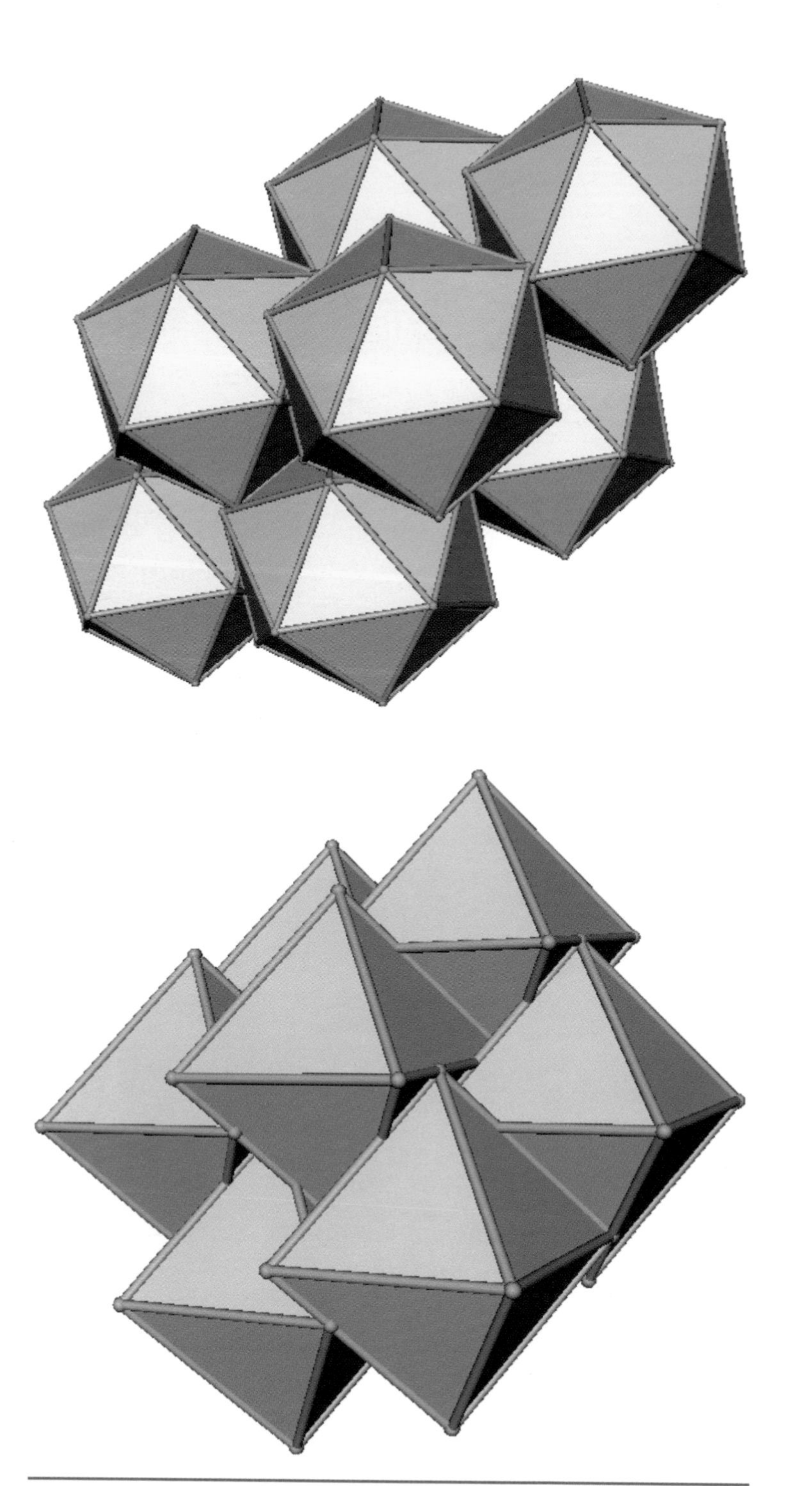

**Figure 7.** *The densest known packings of icosahedra (top) and octahedra (bottom). Both are lattice packings. Torquato and Jiao have conjectured that the optimal packing of centrally symmetric Platonic or Archimedean solids will always be a lattice. (Figure courtesy of Salvatore Torquato. Reprinted figures 10 and 11 with permission from Phys. Rev. E 80, 041104 (2009). ©2009 by the American Physical Society.)*

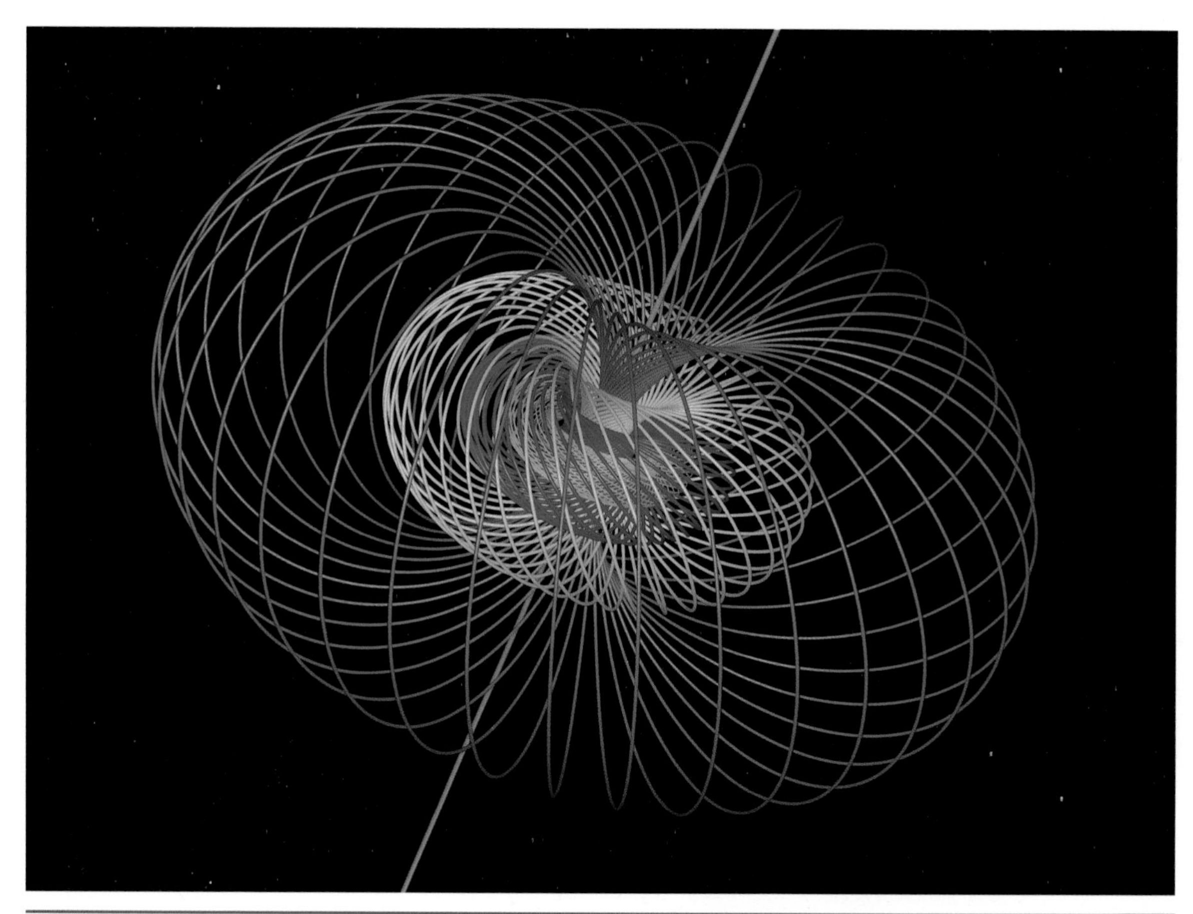

***Linked Pre-images.*** *The Hopf fibration. Every loop in the figure (including the red "loop" that passes through a point at infinity) is linked with every other loop, in such a way that they cannot be separated without breaking one of the loops. The Hopf map takes each of the loops in the Hopf fibration to a different point on an ordinary, 2-dimensional sphere. (Image courtesy of Jos Leys, www.josleys.com.)*

# As One Heroic Age Ends, a New One Begins

THE 1950S AND EARLY 1960S WERE A HEROIC ERA in high-dimensional topology (roughly defined as the study of spaces of more than four dimensions). Early in that period, John Milnor constructed 7-dimensional "exotic spheres," which are identical to normal spheres from the viewpoint of continuous topology, but different from the viewpoint of smooth topology. Later, Michel Kervaire discovered a 10-dimensional continuous space that could not be smoothed. These developments sprang out of a rich new algebraic approach to topology, pioneered by René Thom and others, which enabled topologists to calculate the properties of such seemingly inconceivable spaces.

However, one item of business was left unfinished by this heroic decade: a complete census of all exotic spheres. In 2009, three mathematicians finished this job—except for one obstinate holdout. In the wild and untamed territory of 126 dimensions, the "census takers" only know the number of exotic spheres within a factor of 2.

**John Milnor.** *(Photo courtesy of the Archives of the Mathematisches Forschungsinstitut Oberwolfach.)*

The formal name of the problem solved by Mike Hill of the University of Virginia, Michael Hopkins of Harvard University, and Doug Ravenel of the University of Rochester, is the "Kervaire invariant one" problem. The Kervaire invariant codified the remaining ambiguity left in the classification of exotic spheres.

An $n$-dimensional manifold is a space glued together from copies of $n$-dimensional Euclidean space. Manifolds that can be inserted into a higher-dimensional Euclidean space with a "tube" around them are said to be "framed." (Not all manifolds are framed, and this far-from-obvious fact is itself a discovery from the heroic era.) The Kervaire invariant is a number that can be assigned to any framed $n$-dimensional manifold, and which takes the value either zero or one. The issue was whether any space could actually have a Kervaire invariant equal to one. The only known examples with Kervaire invariant one had been found in dimensions 2 (see Figure 1, next page), 6, 14, 30, and 62. It was also known that any further examples would have to come in dimensions 2 less than a power of 2. Thus 126 would be the next candidate, followed by 254 and 510 and so forth.

"When I was in graduate school, a month didn't go by when I didn't hear somebody talking about the Kervaire invariant one problem," Ravenel says. "Numerous attempts were made to solve it in the 1970s, and they all failed. After a while, it receded into the background." Curiously, the conventional wisdom in the 1970s was that the desired spaces should exist. The discovery of the dimension-30 example and the dimension-62 example (by Michael Barratt, John Jones, and Mark Mahowald in 1984) gave topologists some hope. Ravenel and his colleagues have now dashed this hope. In dimension 254 and beyond, the weird spaces do not exist. Only in dimension 126 is there still some uncertainty, at least for the time being. "Give it five years, and I think it will be resolved one way or the other," says Victor Snaith of the University of Sheffield.

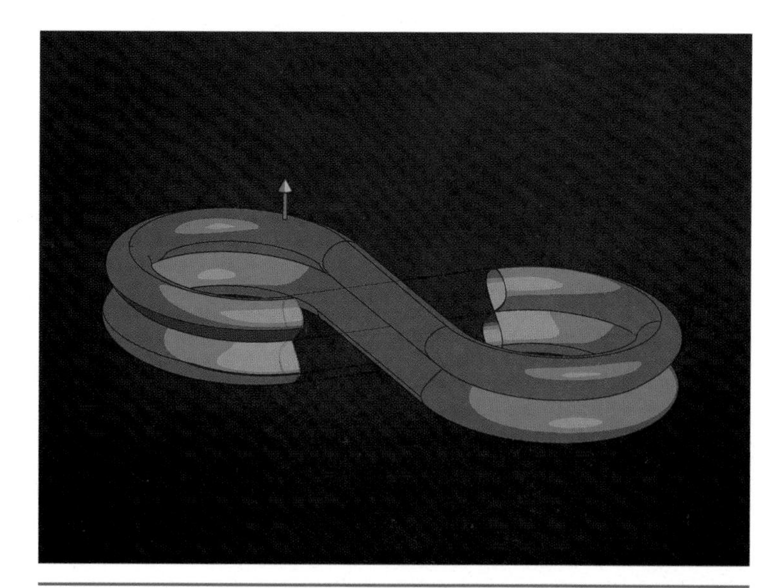

**Figure 1.** *The first example of a framed manifold with Kervaire invariant 1. The manifold is a 2-dimensional torus. In this 3-dimensional rendering, the torus intersects itself; notice that all of the vertical cross sections resemble figure-8s, and the horizontal cross sections likewise. In a correct 4-dimensional view, the intersection points of the figure-8s would separate, and both the horizontal and vertical figure-8s would be framed in the manners shown in Figure 3. Besides this 2-manifold with Kervaire invariant 1, examples have also been found with dimensions 6, 14, 30, and 62 (all 2 less than a power of 2). Uncertainty still remains about dimension 126, but it has now been proved that no such manifolds exist with dimension 254 or higher. (Figure courtesy of Michael J. Hopkins.)*

## The Shot Heard Round the World

John Milnor was scarcely out of graduate school when, in 1956, he fired the shot heard around the world of topology: his proof of the existence of exotic spheres. He won the Fields Medal for his work six years later, making him the second-youngest American ever to win the most prestigious award in mathematics.

A common description of topology is "rubber-sheet geometry." An even more common description (and joke) is that topologists cannot tell their coffee cups apart from their doughnuts. If you start with a coffee cup and expand the handle while shrinking the cup, you will eventually get a doughnut, or torus. To topologists, two manifolds are considered the same, or "homeomorphic," if one can be pulled and stretched and squashed—but not ripped apart or sewn together—until it looks like the other.

However, the "rubber sheet" description conceals the fact that topology takes on a different flavor, depending on whether the manifolds and the functions that map one to another are smooth or just continuous. Smooth manifolds are put together using the functions of calculus; they have no corners or creases, and the maps between them are sinuous and flowing. Continuous manifolds, on the other hand, are glued together piece by piece and may have corners and edges. A continuous manifold could be compared (roughly) to a cubist painting.

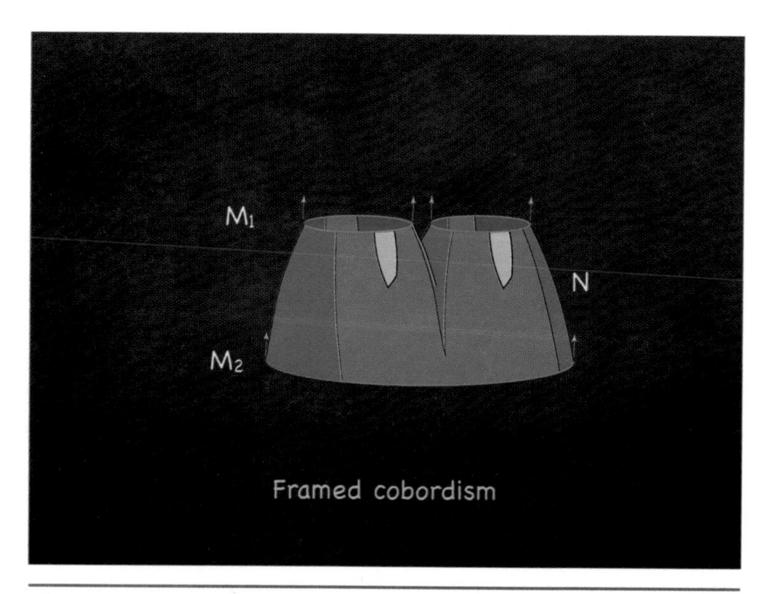


**Figure 2.** *A pair of circles (manifold $M_1$, top) is cobordant to a single circle (manifold $M_2$, bottom) because they form the common boundary of a pair of pants (the blue 2-dimensional manifold, N). (Figure courtesy of Michael J. Hopkins.)*

Until 1956, the difference between continuous and differential topology seemed to be more one of style than substance. For instance, 3-dimensional manifolds can be given a smooth structure, and any homeomorphism between them can be made smooth. But Milnor made it clear that the two schools of topology diverge in higher dimensions. It is possible to find a "cubist" rendering of the 7-dimensional sphere that is so violently distorted that no smooth deformation can undo the damage.

Milnor detected the difference between 7-dimensional spheres by using a global invariant of 8-dimensional manifolds, called the signature. If a smooth, framed 8-dimensional manifold has a smooth sphere as its boundary, then its signature is a multiple of 224. But Milnor constructed a family of 8-dimensional manifolds whose signatures were multiples of 8. Their boundaries, he showed, could be deformed to a sphere by a continuous map. But they could not be smooth spheres because the signatures were wrong. In fact, he concluded, there were 28 different exotic spheres, the boundaries of framed manifolds with signatures 0, 8, 16, 24, and so on up to 216. It was like showing that the Loch Ness monster exists—and that, furthermore, there are exactly 28 of them.

## More Antecedents

Milnor's discovery might have remained just a curiosity, one of the "pathological" examples that topology books are full of, but it turned out to be a crucial cog in the giant machine of algebraic topology.

Besides homeomorphism (the relation that makes a coffee cup equivalent to a doughnut), manifolds can also be related by cobordism. Two manifolds, $X$ and $Y$, are cobordant if a third manifold $M$ (necessarily of one dimension higher) "joins" $X$ to $Y$, that is, the boundary of $M$ is the union of $X$ and $Y$. Thus, for example, two circles are cobordant to a single circle; the proof is a pair of pants (see Figure 2). The waist is the single circle, and the hems of the legs are the other two circles.

The idea of cobordism becomes more complicated, however, when the manifolds are framed. In this case you attach a vector at each point of $X$ and $Y$, pointing perpendicular to the manifold at that point. These vectors enable you to prescribe which side of $X$ and $Y$ you want the bounding manifold to lie on.

On the circle, as shown in Figure 3, there are two distinct framings. One of them makes the circle into the boundary of a disk, with the framing arrow pointing outward. In the other version the circle is given a twist, making it into a figure-8, but the framing is *not* given a twist. If we ignored the framing, the circle would still bound a surface shaped like a potato chip (which is homeomorphic to a disk). However, the outward-pointing normal would twist once around. Thus the figure-8 *with the indicated framing* is not a boundary of the potato chip, and there is in fact no way to make it into the boundary of any disk, and no way to make it cobordant to a regular, untwisted circle.

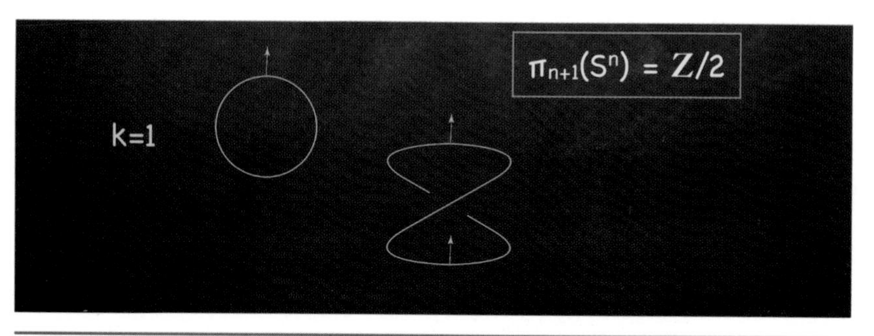


**Figure 3.** *The two framings of a circle. In the standard framing, the circle bounds a disk, and the framing (red arrow) points outward at every point. In the nonstandard framing, the circle is given a twist but the framing points up at every point. This framed curve does not bound a 2-dimensional framed manifold. The torus in Figure 1 is defined as a product of two of these twisted circles. (Figure courtesy of Michael J. Hopkins.)*

One of the original points of framed cobordism was to make it possible to do "surgery" on manifolds. In the 1930s, the Russian mathematician Lev Pontryagin thought that any 2-dimensional surface could be reduced to a sphere by repeatedly cutting out loops and sewing disks over them, preserving the framing at each stage. Figure 4 shows how this procedure might work for a torus: you cut it around a line of longitude, sew disks on both sides, and get a surface that (in rubber-sheet geometry) looks like a sphere. You could also do the same procedure with a line of latitude.

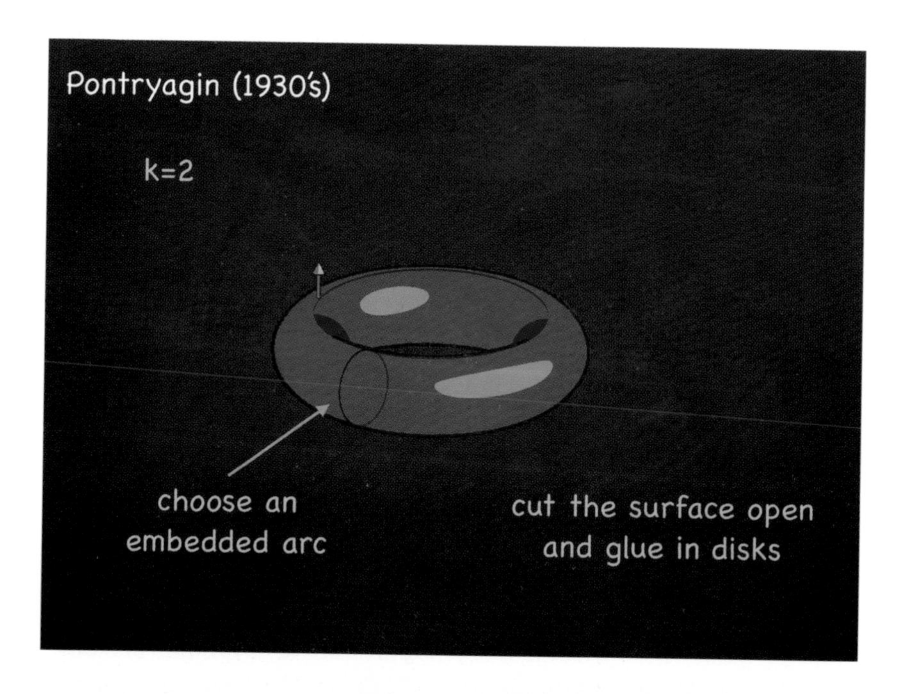


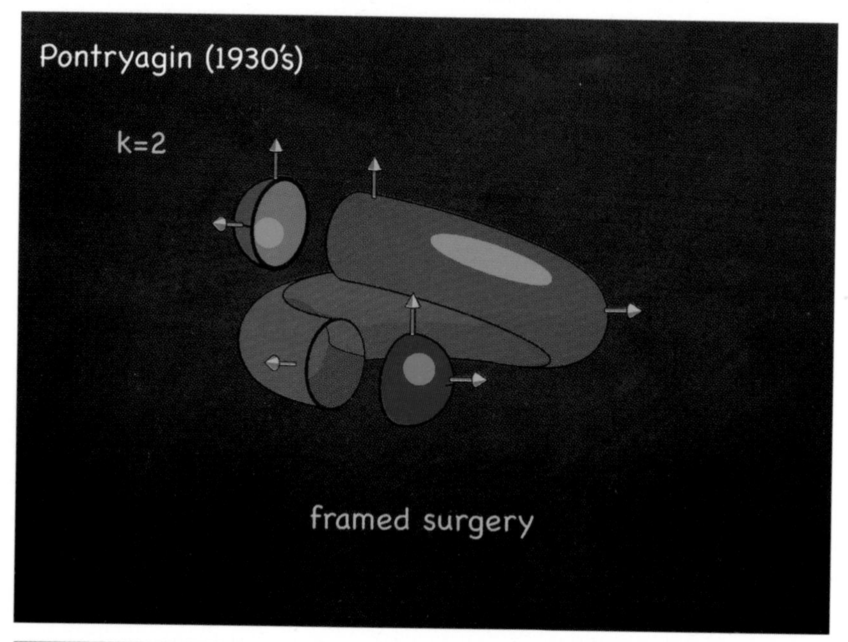


**Figure 4.** *Pontryagin's flawed "proof" that any 2-dimensional manifold is cobordant to a sphere. He assumed that it would always be possible to cut the manifold along a loop and sew a cap onto the exposed cut. But as Figure 1 shows, on a manifold of Kervaire invariant 1 this procedure does not work. Both the horizontal and vertical cross sections are twisted as in Figure 3, and no cap can be sewn on to them. (Figures courtesy of Michael J. Hopkins.)*

**However, three funny things happen in this mad dash to higher dimensions.**

But Pontryagin made one flawed assumption. He assumed that you could always find a framed circle within your manifold that bounds a disk. The framed torus shown in Figure 1 (page 120) proves that he was wrong! Notice that the lines of longitude are circles with a figure-8 framing. So are the lines of latitude. In fact, any closed loop you can draw on this torus will have a framing that makes it impossible to sew a disk onto the loop. This torus is a framed 1-manifold with Kervaire invariant 1. Intuitively, the Kervaire invariant measures the extent of Pontryagin's failure. In dimensions 2, 6, 10, etc., a framed manifold can be converted to a sphere precisely when the Kervaire invariant is zero.

A more precise explanation of the Kervaire invariant involves a concept called homotopy. Homotopies are like lassos that topologists use to detect nontrivial features of a space. For example, you can detect the hole in an inner tube by wrapping a circle around a latitude or a longitude. There is no way to shrink this circle down to a point while keeping it on the surface of the inner tube. Thus it is a nontrivial element of the first homotopy group of the inner tube. In general, the $k$-th homotopy group of a space records all the nontrivial ways of wrapping a $k$-dimensional lasso around it.

If you try to lasso a sphere with a smaller-dimensional lasso, you will never succeed; that is, if $k$ is less than $n$, then the $k$-th homotopy group of the $n$-dimensional sphere (written as $\pi_k(S^n)$) is "trivial." Intuitively, a lower-dimensional sphere inside a higher-dimensional sphere can always be shrunk down to a point because there is no neck for it to get caught on.

You might think that the same thing is true for mappings from a higher-dimensional sphere to a lower-dimensional one. However, human intuition about higher-dimensional objects is pretty poor, and this is one example. A generation before Milnor, Heinz Hopf had discovered that a three-dimensional sphere, $S^3$ (the boundary of a ball in four-space) can be mapped into the ordinary two-dimensional sphere, $S^2$, in a non-trivial way.

The Hopf map can be defined agebraically as follows. Given a point on the sphere $S^3$, you think of it as an ordered pair of complex numbers $(u, v)$, with $|u|^2 + |v|^2 = 1$. You map it to the number $v/u$ in the complex plane. (If $u = 0$, you map it to a "point at infinity.") Finally, you map the plane to $S^2$ by stereographic projection, with the "point at infinity" going to the north pole.

In the resulting map from $S^3$ to $S^2$, every point $z$ in $S^2$ is the image of a circle of points in $S^3$. If we let $z$ move around a loop in $S^2$, its predecessor circles in $S^3$ will trace out a torus, in such a way that each circle *twists* one time as it goes around the torus. This twist is key; it means that the pre-images (or "fibers") of any two points $z$ and $z'$ are always linked. Just as you can't pull a pair of linked handcuffs apart, you can't pull the circles apart, and this is the topological obstruction that prevents the Hopf map from being trivial. Even after 75 years, the geometry of this map is still beautiful to behold (See Figure **Linked Pre-images**, page 118).

Note that Hopf's geometric construction depends on an algebraic trick—the fact that we can divide complex numbers. Similarly, it is possible to divide quadruples of numbers (quaternions) and octuples (octonions), and in this way Hopf also constructed non-trivial maps from $S^7$ to $S^4$ and from $S^{15}$ to $S^8$.

However, three funny things happen in this mad dash to higher dimensions. First, if the dimension of the "target" sphere is increased, and the dimension of the "lassoing" sphere is increased by the *same amount*, the invariants start melting away. Formerly non-trivial maps have a way of becoming trivial. For example, if we bump up the dimension of the Hopf map by one and look at maps from $S^4$ to $S^3$, we discover that the preimages of two distinct points cannot link because circles in 4-dimensional space can always be disentangled. The group $\pi_4(S^3)$ turns out to have only two lassos, rather than infinitely many. We could call the two surviving homotopy classes "twisted" and "untwisted"—just like the two framings of a circle. As it turns out, this is no coincidence!

Second, after a certain point the homotopy groups *always* become finite, and thus they eventually stabilize. For example, $\pi_{n+1}(S^n)$ stabilizes at two homotopy classes. The "stable homotopy" group contains all the homotopy classes that are left after the ephemeral ones melt away. Topologists denote this group as $\pi_1^S$ or $\pi_1(S^0)$. Similarly, $\pi_2^S$ contains all the homotopy classes that are left in $\pi_{n+2}(S^n)$ after $n$ gets sufficiently large. This group also contains two elements, and it is no accident that they correspond to the two framings of a torus (the normal one and the twisted one seen in Figure 1).

The computation of the stable homotopy groups of spheres has not yet been finished, and remains an active field of research. (See Figure 5) It is something like a periodic table for topologists, and the solution of the Kervaire invariant problem has reduced the census of exotic spheres to information contained within this "periodic table."

| $k$ | 1 | 2 | 3 | 4 | 5 | 6 | 7 | 8 | 9 | 10 |
|---|---|---|---|---|---|---|---|---|---|---|
| $\pi_k^S$ | $\mathbf{Z}/2$ | $\mathbf{Z}/2$ | $\mathbf{Z}/24$ | 0 | 0 | $\mathbf{Z}/2$ | $\mathbf{Z}/240$ | $(\mathbf{Z}/2)^2$ | $(\mathbf{Z}/2)^3$ | $\mathbf{Z}/6$ |
| $\Theta_k$ | 0 | 0 | 0 | ? | 0 | 0 | $\mathbf{Z}/28$ | $\mathbf{Z}/2$ | $(\mathbf{Z}/2)^3$ | $\mathbf{Z}/6$ |

| $k$ | 11 | 12 | 13 | 14 | 15 | 16 | 17 | 18 | 19 | 20 |
|---|---|---|---|---|---|---|---|---|---|---|
| $\pi_k^S$ | $\mathbf{Z}/504$ | 0 | $\mathbf{Z}/3$ | $(\mathbf{Z}/2)^2$ | $\mathbf{Z}/480 \oplus \mathbf{Z}/2$ | $(\mathbf{Z}/2)^2$ | $(\mathbf{Z}/2)^4$ | $\mathbf{Z}/8 \oplus \mathbf{Z}/2$ | $\mathbf{Z}/264 \oplus \mathbf{Z}/2$ | $\mathbf{Z}/24$ |
| $\Theta_k$ | $\mathbf{Z}/992$ | 0 | $\mathbf{Z}/3$ | $\mathbf{Z}/2$ | $\mathbf{Z}/8218 \oplus \mathbf{Z}/2$ | $(\mathbf{Z}/2)^2$ | $(\mathbf{Z}/2)^4$ | $\mathbf{Z}/8 \oplus \mathbf{Z}/2$ | $\mathbf{Z}/130816 \oplus \mathbf{Z}/2$ | $\mathbf{Z}/24$ |

**Figure 5.** *The k-th stable homotopy groups of spheres ($\pi_k^S$) and the group of exotic differentiable structures on spheres ($\theta_k$) for $k \le 20$. Milnor's first "shot heard round the world" was the discovery that $\theta_7 \neq 0$ (in fact, there are exactly 28 exotic 7-spheres). René Thom's construction sets up a precise correspondence between these two groups in every dimension except $k = 2, 6, 14, 30, 62, \ldots$ In those dimensions, Thom's construction left an ambiguity due to the possible existence of manifolds with Kervaire invariant 1.*

Another interesting story concerns the *integer-valued* Hopf invariant—in other words, the *unstable* part of the homotopy group $\pi_{n+k}(S^n)$. As noted above, Hopf found interesting elements of $\pi_3(S^2)$, $\pi_7(S^4)$, and $\pi_{15}(S^8)$. His construction uses the

**Michel Kervaire.** *(Photo used in Andrew Ranicki's presentation (http://www.maths.ed.ac.uk/~aar/slides/kervaire.pdf) at the Kervaire Memorial Symposium, Geneva, February, 2009.)*

fact that it is possible to divide ordered pairs, quadruples, and octuples of numbers. The next step would seem to be $n = 16$. Is there some way to divide vectors in 16-dimensional space? Remarkably, the answer is no.

In 1898 (before the discipline of topology even existed), Adolf Hurwitz had proved that there is no "normed division algebra" over the real numbers in any dimension other than 1, 2, 4, or 8. Hurwitz's theorem placed one extra condition on the operation of division—it has to preserve norms (so that $|u/v| = |u|/|v|$. In 1958, Kervaire and Milnor showed that the extra condition is unnecessary; there are no division algebras of *any* kind except in dimensions 2, 4, and 8. This still left open the possibility that there might be some other ingenious way to define the Hopf invariant in higher dimensions. But in 1960, using a new technique called the Adams spectral sequence, Frank Adams showed that this, too, was impossible. The Hopf invariant dies after dimension 8.

This problem—the "Hopf invariant one" problem—would be a direct antecedent of the Kervaire invariant one problem. Its resolution was also quite similar; the invariant exists only in a finite number of exceptional cases, all related to powers of 2.

## How It All Fits Together

Cobordisms, homotopy groups of spheres, and exotic spheres are all tied together by a beautiful construction discovered by René Thom in 1954—around the beginning of the heroic decade. (Thom, like Milnor, received a Fields Medal for his work.) Pontryagin had already come up with the same idea, although he came about it from the opposite direction as Thom and is thus regarded as a co-inventor.

Pontryagin and Thom realized that *framed cobordism* and *stable homotopy* are essentially the same concept. Here is why. Suppose you have any framed $k$-dimensional manifold $M$, embedded in $(n+k)$-dimensional space, with a little tube $T$ about it. If you slice across the tube, you get a small $n$-dimensional ball that has a preferred coordinate frame, which comes from the framing of $M$. Thus each point $x$ in the tube has a unique "$M$-coordinate" (the point $m$ in $M$ closest to $x$) and an "$\mathbf{R}^n$-coordinate" (the vector that points from $m$ to $x$).

The tube $T$ can be used to create a mapping from $S^{n+k}$ to $S^n$, and therefore an element of $\pi_{n+k}(S^n)$, as follows. Map every point inside the tube $T$ to its " $\mathbf{R}^n$-coordinate." The image of each slice across the tube will be a tiny ball in $\mathbf{R}^n$. Explode this ball so that its boundary goes out to infinity. Notice that, as a result, every point on the boundary of the tube $T$ is getting mapped to the "point at infinity." Map every point outside the tube $T$ (including the "point at infinity" in $\mathbf{R}^{n+k}$) to this same "point at infinity" in $\mathbf{R}^n$. Finally, use stereographic projection to curl up $\mathbf{R}^n$, together with its "point at infinity," into a sphere $S^n$.

For example, when $M$ is a circle in 3-dimensional space, $T$ is a solid torus. If $M$ is twisted, then the Thom-Pontryagin construction results in a map that is homotopic to the Hopf map. An untwisted framing would lead to a homotopically trivial map from $S^3$ to $S^2$. This is why the "twisted-untwisted" dichotomy of Hopf maps is connected to the two framings of a circle.

Thom proved that this Hopf example is completely typical. To be more precise, whenever two framed manifolds are cobordant, their Thom-Pontryagin maps are homotopic. Whenever they are not cobordant, their Thom-Pontryagin maps are different. And every stable homotopy class in $\pi_{n+k}(S^n)$ does actually come about via this construction.

Now, the plot starts to thicken when we marry Thom's isomorphism to Milnor's exotic spheres. We can ask three questions:

1). In the correspondence between stable homotopy classes and framed manifolds, which homotopy classes come from exotic spheres?

2). Which homotopy classes do *not* come from exotic spheres?

3). And which exotic spheres correspond to trivial homotopy classes? (This happens when the exotic sphere is cobordant to a regular sphere—in other words, it is the boundary of an ordinary $(k+1)$-dimensional manifold.)

In a truly ideal world, the answers to these three questions would be "All of them," "None of them," and "None of them," respectively. If that were the case, there would be a perfect match between the groups of exotic spheres and the stable homotopy groups. As you can see from Figure 5, the match is not perfect—but it is pretty close. In dimensions that are 1 greater than a multiple of 4, the match is perfect. In dimensions 3 greater than a multiple of 4, the match is not precise but it is at least roughly understood. The stable homotopy groups and the group of exotic spheres correspond, in a known way, to the denominator and the numerator of a sequence of numbers called the Bernoulli numbers. Jakob Bernoulli, who discovered these numbers in 1713, could never have dreamed the use that they would one day be put to.

And then there are the dimensions 2, 6, 10, 14, etc., which are 2 greater than a multiple of 4. These are precisely the dimensions where the Kervaire invariant comes into play. Once again, Kervaire proved that the three questions posed above are closely related. If a manifold of Kervaire invariant one exists, then the answer to the three questions above are: Half of them, half of them, and none of them. If no manifold of Kervaire invariant one exists, then the answers are: All of them, none of them, and half of them. If only we knew which set of answers was the correct one! The subject was stuck in this limbo for nearly 40 years.

**. . . the plot starts to thicken when we marry Thom's isomorphism to Milnor's exotic spheres.**

### The Elusive Kervaire Invariant One Monster

At this point it is probably a good idea to sort out what these questions really mean. If there is a homotopy class that does *not* come from an exotic sphere, it must come from some other manifold by the Thom-Pontryagin construction. This means that we have found a manifold that is not cobordant to a sphere, a manifold of Kervaire invariant one. If there is an exotic sphere that does *not* give a non-trivial homotopy class, then by Thom's theorem it must be cobordant to a regular sphere. Kervaire's theorem tells us that in every dimension 2, 6, 10, etc., one of these unusual creatures exists and the other does not. Now, thanks to the theorem of Hill, Hopkins, and Ravenel, we know which one is truly rare: the Kervaire Invariant One Monster.

**Photo of Mike Hill, Doug Ravenel, and Michael Hopkins.** *(Photo courtesy of Bill Browder).*

In 1969, William Browder pushed these monsters a long way toward extinction. He showed they could only exist in dimensions 2, 6, 14, 30, etc.—all numbers that are 2 less than a power of 2. The first three Kervaire Invariant One Monsters can be produced by "squaring" the 1-sphere, 3-sphere, and 7-sphere respectively. Figure 1, page 4 shows the first one, a product of two copies of the twisted circle.

For dimension 30, squaring the 15-sphere doesn't work, for reasons directly related to the nonexistence of division algebras in dimension 16. But several Monsters have been captured, including a pretty one discovered in 1980 by Snaith and Jorgen Tornehave that involves a fourfold product of $S^7$'s with a 5-holed torus. (Note that the dimensions add. The fourfold product of $S^7$'s has dimension $4 \times 7 = 28$, and multiplying this object by the 5-holed torus, which is a 2-dimensional surface, brings the dimension up to $4 \times 7 + 2 = 30$.) "We thought this was so easy that we must have the general method!" Snaith said. "Unfortunately, we spent the rest of the month proving to ourselves that our method fizzled out precisely at dimension 30." The existence of 62-dimensional Kervaire Invariant One Monsters was proved first in 1984, but it took a book-length computer calculation by Stan Kochman of York University in 1990 to produce a definitive construction.

After so many years with so little progress, Ravenel, Hill and Hopkins were not actually expecting to solve the Kervaire invariant one problem when they started working together in 2008. They actually weren't working on that problem at all. "It's as if we were hiking in the Alps and suddenly found a shortcut to Mount Everest," Ravenel says. They were working on generalizations of the Adams spectral sequence, which Frank Adams had used to solve the Hopf invariant one problem.

Even for mathematicians, the concept of a spectral sequence is rather abstruse. Ravenel describes it as a way of writing down everything that could possibly be a map between spheres. Like a human genome, it contains a lot of "junk DNA"—maps that don't really exist—and so the trick, for a topologist, is to filter out the real stuff from the junk.

The original Adams sequence is too difficult to calculate in full. But topologists have found ways to alter the definition to throw out some of the junk, along with some of the good stuff too, but in the process make the whole thing easier to calculate. The trick then is to make sure that you haven't tossed out the "good stuff" that you actually wanted to keep.

In recent years, Hill and Hopkins became interested in a particular modification of the Adams spectral sequence (called the Adams-Novikov sequence) that uses what they call topological modular forms. This sequence was easier to calculate because it has deep connections with number theory and algebraic geometry, and it also had two very desirable properties. First, it is periodic (meaning that the same maps appeared every 8 dimensions or so), and second, it has a spectral gap. The gap is a set of dimensions, usually negative, where the homotopy groups are zero. If the concept of negative dimensions flummoxes you, just relax and remember that the spectral sequence is an abstract algebraic construct. "We don't even know a good geometric description of the thing whose homotopy groups we're computing," says Ravenel.

It took them a few months of tinkering, but the three mathematicians finally discovered a spectrum $\Omega$ with a period of 256 and a spectral gap at dimension $-2$, and which moreover was capable of detecting the Kervaire invariant. (In other words, it didn't toss out the good stuff.) The spectral gap means that $\pi_{-2}(\Omega) = 0$. The periodicity then implies that $\pi_{254}(\Omega) = 0$. If some 254-dimensional manifold $M$ existed with Kervaire invariant one, it would produce a corresponding non-trivial element of the stable homotopy group $\pi_{254}^S$, by the Pontryagin-Thom construction. That non-trivial element would map to a non-trivial element of $\pi_{254}(\Omega)$. But that group has no non-trivial elements, and so the supposed manifold $M$ does not exist.

Unlike some nonexistence theorems in mathematics, this one is not a dead end. From the point of view of homotopy theory, the fact that the Kervaire invariant is unstable means that something unexpected must happen just before it disappears. "Certain homotopy theorists had developed a very pleasant description of the homotopy groups of spheres assuming the Kervaire invariant was present," Ravenel says. "We need a new vision of what the world looks like."

Finally, of course, the Kervaire invariant one problem may be just the beginning of what topologists will learn from the new machinery of topological modular forms and the Adams-Novikov sequence. As Ravenel says, "We feel we've got a tiger by the tail."

**From the point of view of homotopy theory, the fact that the Kervaire invariant is unstable means that something unexpected must happen just before it disappears**